BUILDING SKIN & DETAILS

建筑表皮细部结构

高巍 编　贺丽 译

辽宁科学技术出版社

Contents
目 录

Glass Façade 玻璃立面

4 *Ramboll Head Office*
Ramboll公司总部

14 *San Joaquin County Administration Building*
美国加州圣华金郡行政办公楼

22 *Westraven*
Westraven办公楼

32 *Menara Perak*
Menara Perak办公楼

40 *Palestra*
Palestra办公大楼

48 *Mediacomplex*
Mediacomplex办公大楼

58 *UNIQA Vital Business Centre*
优尼卡保险公司重要业务中心

64 *Creative Valley*
创意谷

70 *Basque Health Department Headquarters*
巴斯克卫生署总部

Metal Façade 金属立面

78 *Bermuda Point Office Building*
百慕大顶点办公大楼

86 *Schwandorf*
Schwandorf行政大楼

96 *Nestle Application Group*
雀巢咖啡公司大楼

106 *Cocoon*
蚕茧办公大楼

114 *Perimeter Institute for Research in Theoretical Physics*
加拿大理论物理研究所

122 *Rotermanni Vana Jahuladu ja Uus Jahuladu*
Rotermann Jahuladu办公楼

134 *Steckelhörn 11 - Office Complex*
Steckelhörn 11综合办公大楼

142 *Financial and Commercial Department of Voestalpine Stahl GmbH*
奥钢联斯塔尔有限公司金融与商业部办公大楼

152 *Audenasa*
西班牙Audenasa办公建筑

158 *No. 5 Merchant Square*
帕丁顿商业广场5号

Stone Façade 石材立面

166 *City Hall of the Shanghai Nanhui District*
上海南汇区市政厅

174 *Aspis Bank Headquarters*
Aspis银行总部

Concrete Façade 混凝土立面

182 *Monaco House*
摩纳哥综合办公大楼

190 *Oporto Vodafone Building*
波尔图沃达丰大楼

Timber Façade 木材立面

200 *Maasberg Pavilion in Overloon*
上洛恩少年管教所

208 *Head Office Irdeto*
爱迪德公司总部

Complex Façade 综合立面

218 *Office Building VDAB*
VDAB办公楼

230 *Horizontal Skyscraper - Vanke Centre*
水平摩天楼——万科中心

240 *Swedbank Head Office Building*
瑞典银行总部大楼

248 *BDP Manchester Studio*
BDP曼彻斯特工作室

258 *Cabel Industry*
Cabel Industry 软件公司总部

266 *Advice House*
咨询大厦

274 *Seewurfel, Mixed-Use Regeneration Development*
湖水立方体多功能再生工程

286 *Index*
索引

Ramboll Head Office

Ramboll公司总部

Location: Copenhagen, Denmark
Designers: Dissing+Weitling Architecture
Photographers: Adam Mork
Completion date: 2010

项目地点：丹麦，哥本哈根
设计师：Dissing+Weitling建筑师事务所
摄影师：亚当・莫克
竣工时间：2010年

Award name:
WAN Award Shortlist 2010

获奖名称：
2010年世界建筑新闻奖提名

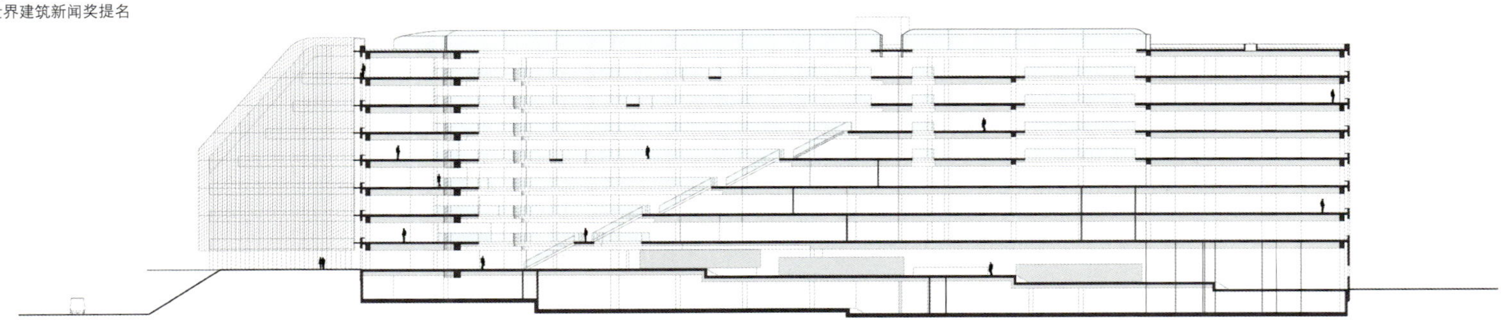

Four years of work have resulted in one of Denmark's largest and most dramatic corporate head quarters, the Ramboll Head Offices, RHO. This 40,000 sq m office is now the daily workplace for 1,800 employees, yet every detail has been designed to reflect and underpin Rambolls encompassing vision of innovation, collaboration and knowledge transfer.

Ramboll requested a new head office that could support a 'one company' understanding. The building should integrate meeting places for formal and informal meetings, and create room for innovation and networking across different fields. Ramboll wanted with the building to communicate a transparent and responsible organisation with a holistic approach.

Dissing+Weitling Architecture has translated this vision in to an architectural concept that draws its inspiration from Barcelona's famous Rambla - the energetic and vibrant main street. The RHO-Rambla unifies the entire building both horizontally and vertically through a series of mezzanine floors and balconies, in effect a massive movement machine that allows staff and visitors to move effortlessly between the eight floors. A generous open stairway lead via semi public spaces and meeting rooms related the central space to the top floor. The Rambla concept makes it surprisingly easy to navigate. The moment the visitor step inside, the building unfolds and reveals its open floors. There is a clear transition from social, common spaces to quiet office pockets. The building even includes a café open to the public.

The building is engineered with a holistic approach leading to long-term and creative results, reducing CO_2 emission and integrating responsible climate friendly solutions. With reduction of energy through optimal daylight solutions, district heating and ground water cooling, reducing cooling needs by up to 85%, the building will have a low consumption of energy. Evaluating energy, transport, water, health and wellbeing, the building archive a rating similar to BREEAM Good.

The building is a result of an open-minded and creative process between users, engineers, architects, investor and contractor. The compact and movement-driven central space holds a unique quality and is a modern and new response to increasing demands for buildings designed for innovation and corporation. With its walkways and curved balconies bustling with life, the interior challenges the rigour of the seamless 5,000 sq m double-skinned glass façade creating the impression of an open, yet refined piece of architecture.

历经四年的打造，丹麦最大、最引人注目的Ramboll公司总部在哥本哈根成功落成。这个占地40,000平方米的办公间完美地成为1,800名员工的办公之家，设计的每一个细节均以彰显并强调Ramboll公司的创新意识、合作精神与知识结构为主旨。

Ramboll公司新总部的设计要求建筑能够体现“一个公司”的理念。建筑能够将正式会议间和非正式会议间完美结合在一起，并为不同领域提供创意和交流的空间。该公司希望新建筑能够采用整体分析的方式有力地彰显出其组织的透明化和可靠性。

Dissing+Weitling建筑事务所巧妙地将这一构想融入到设计理念之中，设计的灵感源自对巴塞罗那著名的兰布拉大街的参考。该建筑通过一系列的阁楼和露台将建筑的水平和垂直方向完美衔接，犹如一个大型运转机器令员工和访客可以自由地在八层的空间中自由穿梭。一个宽敞的开放式楼梯自然地将半公共空间和会议室以及中央空间和顶层衔接在一起。兰布拉大街的设计理念使空间的布局更加清晰、易辨别。访客在迈进空间的一瞬间，即会被精心设计的开放式楼层所震撼和感染。交流、公共空间和办公间之间的过渡清晰、自然。此外，该建筑内还为公众设置了一个自助咖啡厅。

建筑采用整体分析的方式，外观独特，内部具有良好的可持续性，以降低二氧化碳排放量、尽量减少对环境的消极影响为设计的出发点。空间采用优化的能源解决方案，有效利用太阳能，尽量降低对能源的利用，采用集中供暖和地下水冷却的方式，将对制冷的能源消耗量降低85%，尽可能地减少了对能源的消耗。如果从能源、交通、水源、卫生和康乐等角度来对建筑进行评估，该建筑完全达到了建筑研究院环境评估的优质等级。

该建筑是用户、工程师、建筑师、投资者和承包商集思广益、群策群力的结晶。简洁而不失活力的中央空间风格独特，具有浓厚的现代气息，是对创新型办公建筑的一次全新尝试。通道和弧形露台繁忙而充满活力，无缝衔接的5,000平方米双层玻璃表面令建筑分外通透和精致。

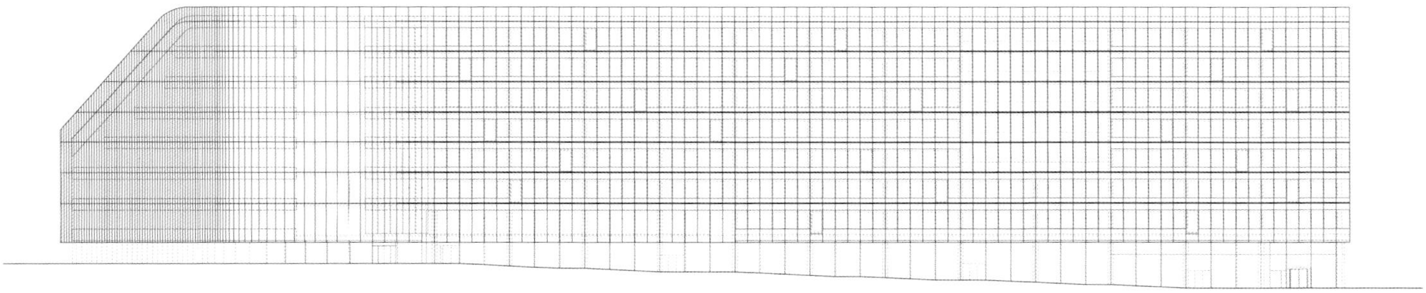

RAMBØLL

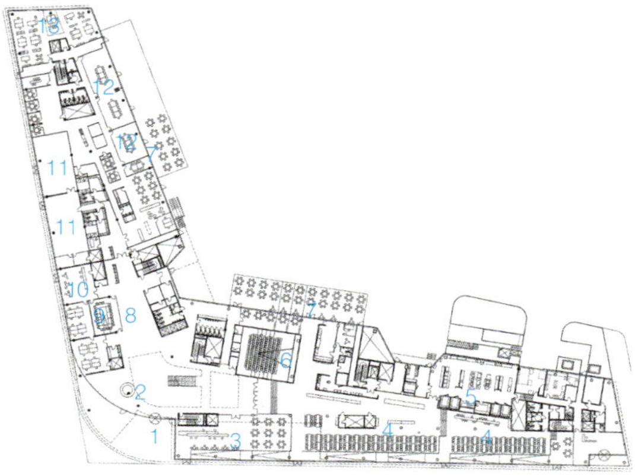

1. Main entrance	1. 主入口
2. Reception	2. 接待处
3. Café	3. 咖啡厅
4. Canteen	4. 餐厅
5. Kitchen	5. 厨房
6. Auditorium	6. 礼堂
7. Terrace	7. 阳台
8. Exhibition	8. 展览室
9. Conference	9. 会议间
10. Lounge	10. 休息室
11. Fitness	11. 健身室
12. Education	12. 教学间
13. Office	13. 办公间

RAMBOLL
CAFE HANNEMANN

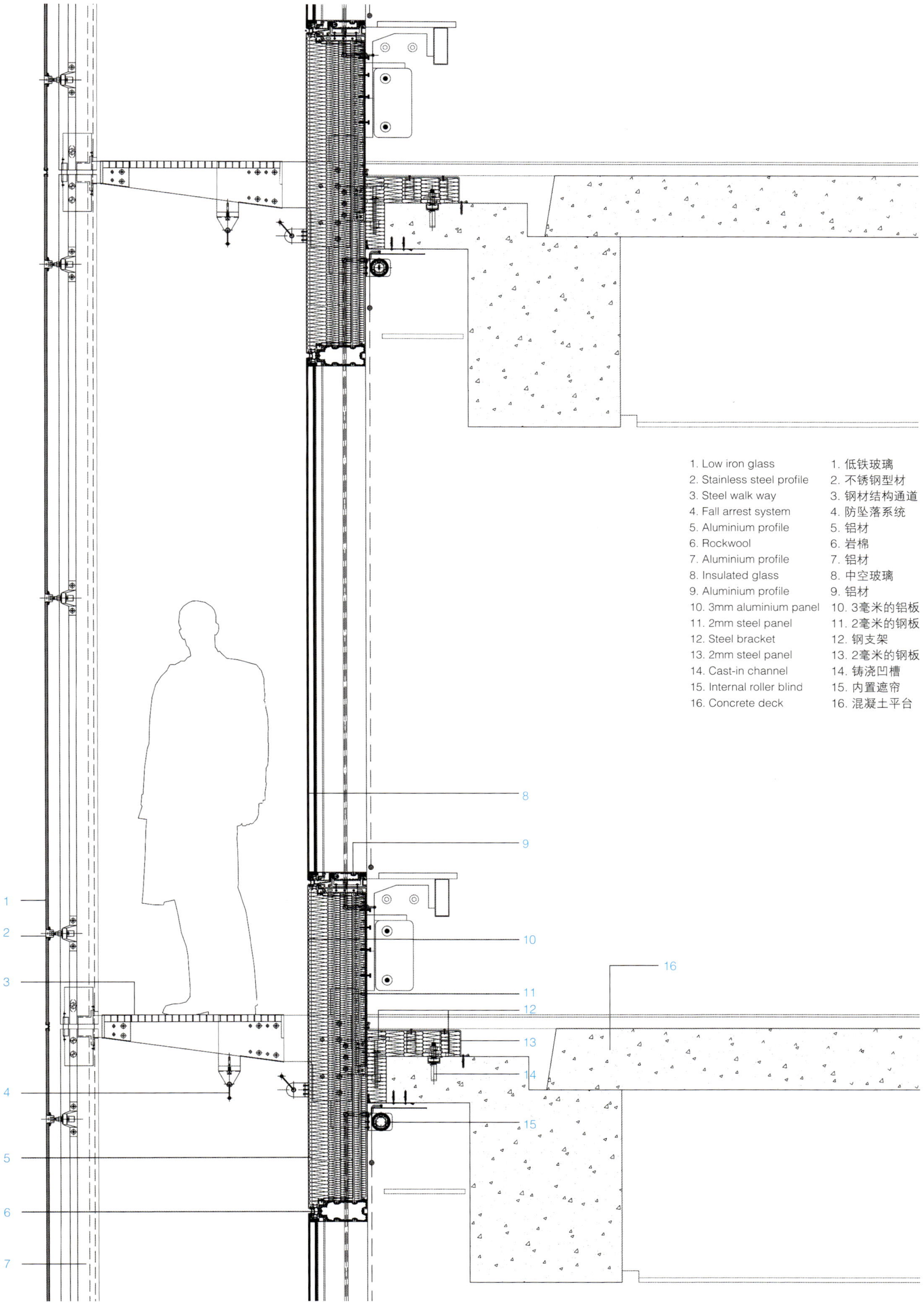

1. Low iron glass	1. 低铁玻璃
2. Stainless steel profile	2. 不锈钢型材
3. Steel walk way	3. 钢材结构通道
4. Fall arrest system	4. 防坠落系统
5. Aluminium profile	5. 铝材
6. Rockwool	6. 岩棉
7. Aluminium profile	7. 铝材
8. Insulated glass	8. 中空玻璃
9. Aluminium profile	9. 铝材
10. 3mm aluminium panel	10. 3毫米的铝板
11. 2mm steel panel	11. 2毫米的钢板
12. Steel bracket	12. 钢支架
13. 2mm steel panel	13. 2毫米的钢板
14. Cast-in channel	14. 铸浇凹槽
15. Internal roller blind	15. 内置遮帘
16. Concrete deck	16. 混凝土平台

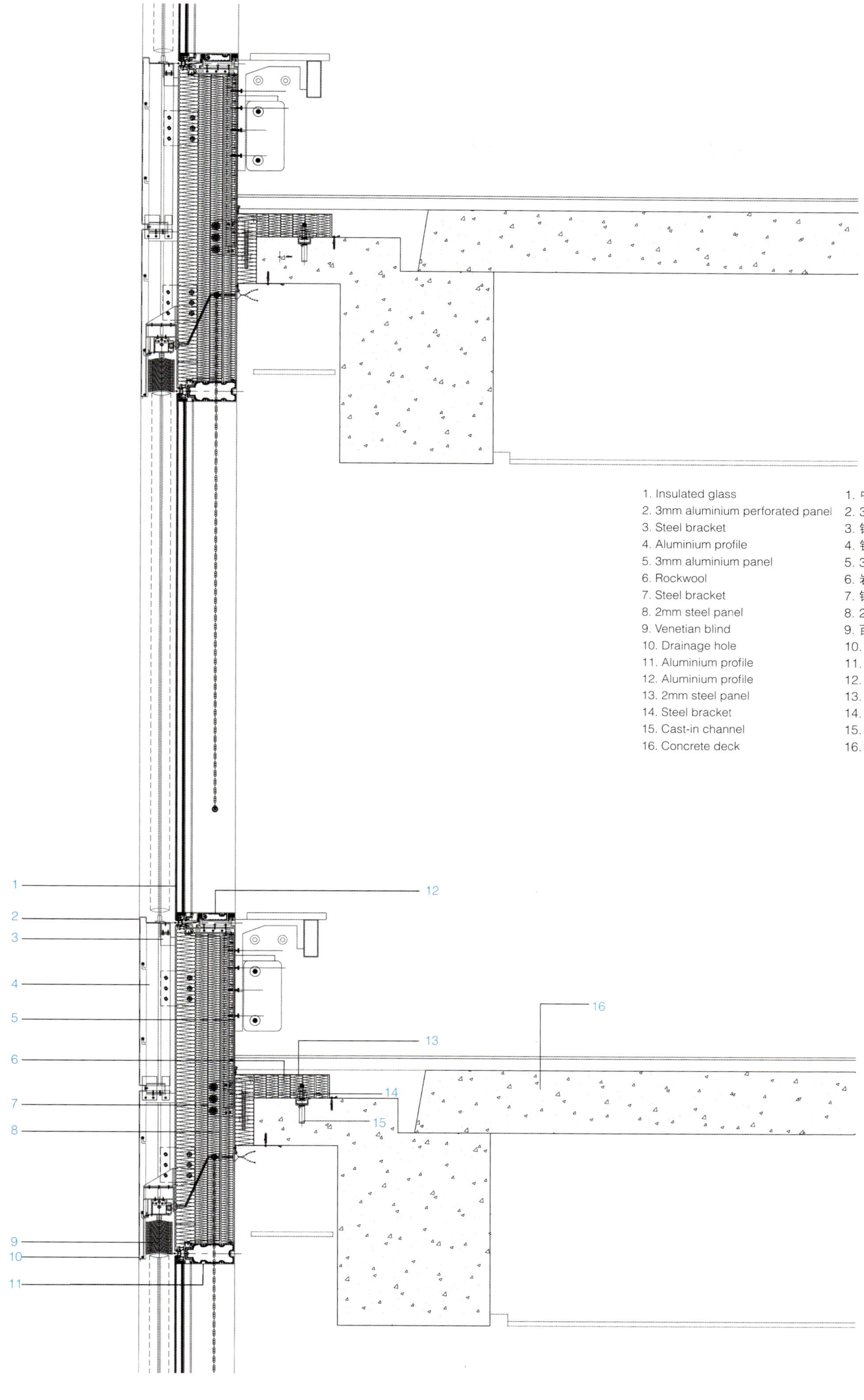

1. Insulated glass
2. 3mm aluminium perforated panel
3. Steel bracket
4. Aluminium profile
5. 3mm aluminium panel
6. Rockwool
7. Steel bracket
8. 2mm steel panel
9. Venetian blind
10. Drainage hole
11. Aluminium profile
12. Aluminium profile
13. 2mm steel panel
14. Steel bracket
15. Cast-in channel
16. Concrete deck

1. 中空玻璃
2. 3毫米穿孔铝板
3. 钢支架
4. 铝材
5. 3毫米铝板
6. 岩棉
7. 钢支架
8. 2毫米的钢板
9. 百叶窗帘
10. 排水孔
11. 铝材
12. 铝材
13. 2毫米的钢板
14. 钢支架
15. 铸浇凹槽
16. 混凝土平台

San Joaquin County Administration Building

美国加州圣华金郡行政办公楼

Location: California, USA
Designers: Fentress Architects
Photographers: Jason A. Knowles © Fentress Architects, Nick Merrick © Hedrich Blessing
Completion date: 2009

项目地点：美国，加利福尼亚州
设计师：Fentress建筑事务所
摄影师：杰森·A·诺雷斯/ Fentress建筑事务所，尼克梅里克/ Hedrich Blessing摄影工作室
竣工时间：2009年

Award name:
2009 Project of the Year Award, American Public Works Association, Sacramento Chapter
2008 Design Excellence Award, Novum Structures

获奖名称：
美国公共工程协会萨克拉门托分会2009年度最杰出项目奖
2008年Novum Structures有限责任公司优秀设计奖

"Democracy means transparent government," says architect Curtis Fentress. "The abundance of glass used for the atrium and Board Chambers literally and symbolically promotes open and accessible government." Representing the transparency and accessibility of government, the building's innovative five-storey, sculptural glass atrium rises from the corner plaza and culminates in a sixth-floor, 200-seat board chamber cantilevered above. Using context to create identity, a signature of Fentress' designs, the atrium was inspired by the natural landscape surrounding San Joaquin County, evoking images of mountains, agriculture and the Stockton Maritime Port. The atrium's glass form recalls the natural beauty of peaked, protruding stone formations found within nearby Yosemite National Park, a tranquil refuge from busy city life. The atrium folds inward at the entrance, welcoming visitors into a dramatic daylit lobby that conveys a grand sense of arrival.

The new Administration Building establishes a forward-looking image and public presence for the San Joaquin County government and downtown Stockton. Consolidating administrative functions and government services for easy access by citisens, the building intends to serve as a catalyst for revitalisation in downtown Stockton. Using design techniques such as scale, rhythm, composition, massing, colour and materials, Fentress mindfully blends the contextual structure with Stockton's historic downtown to create a human scale. Arcades and an entry plaza encourage pedestrian activity and urban vitality. The energy-efficient building utilises recycled and low-VOC materials and is pending LEED Gold Certification by the U.S. Green Building Council.

“民主意味着政府的透明”，建筑设计师柯蒂斯·芬特雷斯这样说道。“通透的玻璃中庭和木板内庭设计从外观和象征意义的角度体现出政府开放、透明的管理。”这个匠心独运的五层办公建筑，充分彰显了政府透明、公开的管理模式。覆有艺术彩雕玻璃的中庭从广场的一角出发一直向上延伸，最终抵达建筑顶端的悬垂木板内庭。这个作为“第六层”的悬垂结构中设置了200个座位。利用周围的环境背景是Fentress建筑事务所的设计特色。该中庭的设计即受到圣华金郡周围的自然景观的启发，与连绵起伏的山脉、蓬勃发展的农业以及斯托克顿码头交相辉映。中庭的玻璃外观彰显了城市净土－约塞米蒂国家公园附近山石的自然之美。中庭在入口处向内凹折，邀请访客进入到充满活力的入口大厅，通透、宽敞的大厅在第一时间向访客传达出大气、高雅的空间感。

新行政办公楼为圣华金郡政府和斯托克顿市区打造了崭新的公众形象。该建筑进一步完善了政府的行政职能和公共服务系统，为斯托克顿市区的振兴注入了无限活力。设计师巧妙地综合运用比例、节奏、构图、概念图、色调和材料等设计手法，将建筑结构与历史悠久的斯托克顿市区特色相结合，从而构建一个合理的人文尺度。拱形游廊和入口广场为行人提供了完美的步行空间，使城市焕发出勃勃生机。此外，该建筑采用可再生和低挥发性有机化合物材料，有效降低能源消耗，提高能源利用率。同时，该建筑正在等待美国绿色建筑协会的金级认证。

JOAQUIN

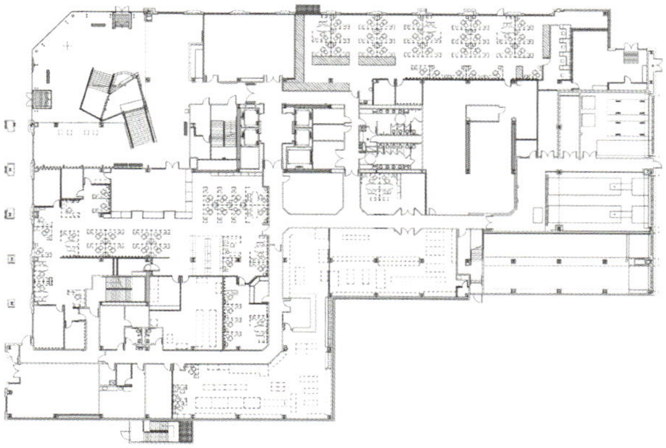

NISTRATION BUILDING

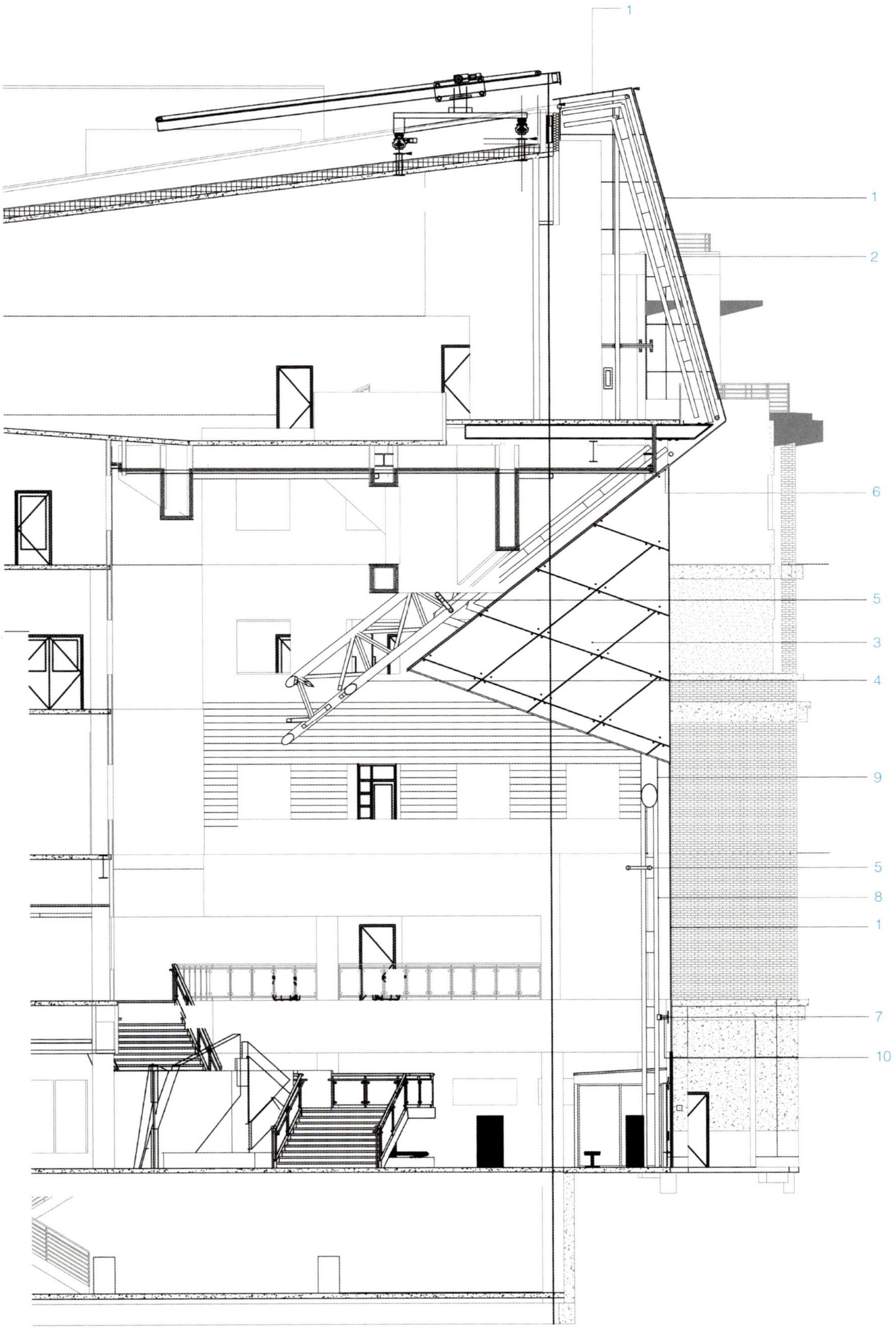

1. Green tinted laminated point-supported glass, 21.5mm
2. Round steel hollow tube, 8.625 inches Diameter
3. Green tinted laminated point-supported glass, 23.5mm
4. Three-dimensional steel truss, 56 inches deep
5. Steel pratt truss, 24 inches deep
6. Silicone expansion gasket
7. Stainless steel glazing arm, 6 inches reach
8. Vertical steel vierendeel truss, 14 inches deep
9. Round steel hollow tube, 16.75 inches Diameter
10. Steel glazing arm support fin

1. 绿色夹层点支式玻璃，21.5毫米
2. 圆钢空心管，8.625英寸直径
3. 绿色夹层点支式玻璃，23.5毫米
4. 三维钢桁架，56英寸高
5. 普拉特钢桁架，24英寸高
6. 硅胶膨胀垫圈
7. 不锈钢玻璃臂，6英寸
8. 空腹桁架竖向钢，14英寸高
9. 圆钢空心管，直径 16.75英寸
10. 玻璃钢臂肘支架

Westraven

Westraven办公楼

Location: Utrecht, The Netherlands
Designers: Cepezed Architects
Photographers: Jannes Linders
Completion date: 2007

项目地点：荷兰，乌得勒支
设计师： Cepezed建筑师事务所
摄影师：雅尼·林德斯
竣工时间：2007年

Award name:
Daylight Award, Dutch Construction Award, Nomination for BNA Building of the Year, Nomination for National Steel Prize 2008
2008 to the Daylight Award
2009 to the Dutch Construction Award
2008 to the Nomination for BNA Building of the Year
The 2009 Best Tall Building Award Europe of the Council on Tall Buildings and Urban Habitat in Chicago
The Rietveld Prize 2009
The Mies van der Rohe Award 2009
The Prime Property Award 2010 for sustainable developments

获奖名称：
日光奖，荷兰建筑奖，荷兰建筑师协会年度建筑提名奖，2008年国家钢铁奖提名
2008年日光奖
2009年荷兰建筑奖
2008年BNA建筑年度提名奖
2009年欧洲高层建筑与城居委员会最佳高层建筑奖
2009年里特维尔德奖
2009年密斯·凡·德罗奖
2010年可持续发展项目最佳性能奖

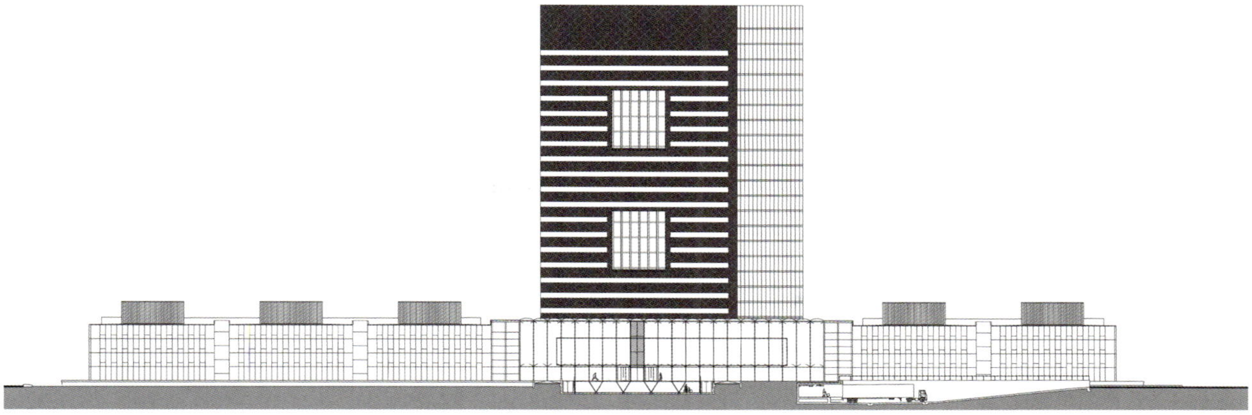

This project is a large-scale combination of an existing building and new construction for various departments of the Ministry of Public Works, situated in the Westraven area between the Amsterdam-Rhine Canal and the junction of the A2 and A12 motorways at Utrecht. The programme covers more than 53,000 metres and consists of office space, conference facilities, a national meeting centre, a communications centre, and the Future Centre for the Ministry of Public Works. More than 2000 people are employed at Westraven.

The existing high-rise construction has been completely renovated and reorganised. An elongated four-storey building has been realised around the foot of the building. Façades made entirely of glass and voids in the floors give a strong impulse to the spatial experience of the high-rise block was rather limited in the original architecture. Atriums, conservatories and inner gardens stimulate unprecedented spatial awareness, and are also important for the orientation of staff and visitors. The various functions of high-rise and low-rise construction are combined at the foot of the tower in spectacular, large, open spaces that invite both a rendezvous and casual encounters. Many of the working areas have been designed as flexible workstations, so that it is possible to work at various places in the building.

Much attention has been devoted to realising perfect equilibrium between low energy consumption and an optimum working climate. For this purpose, an ingenious climate concept has been developed which makes use of thermally active floors. Various types of spaces have been climatised differently, the heat of air flows is recovered at several points and the artificial lighting adapts itself automatically to the colour and intensity of the incidence of daylight.

Furthermore, a revolutionary feature is the innovative second-skin façade of open-weave, teflon-coated glass fibre that offers all-round protection against the wind and makes it possible to work with open windows, even on the top floors. As a result, the staff enjoy great freedom of regulating their own interior climate, which is quite a luxury for a high-rise construction.

该项目是一个新旧建筑结合的大型综合性建筑,是荷兰公共工程部各部门的办公场所,位于阿姆斯特丹–莱茵运河和A2和A12高速公路在乌得勒支汇合处之间的Westraven地区。建筑占地53,000平方米，内部包括办公空间、会议设施、国家会议中心、通讯中心、公共工程部未来中心，容纳员工超过2000人。

原有的高层建筑经过彻底整修和组建重新焕发出勃勃生机。紧邻高层建筑一栋狭长的四层楼高建筑拔地而起，其立面完全采用玻璃材料，楼层之间的衔接与高层街区的空间体验形成一种强烈的视觉冲击，而这在最初的建筑中是相当受限的。中庭、温室与室内花园激发了前所未有的空间感，帮助员工和访客进行准确的定位。高层建筑与低层建筑的不同功能区在建筑的下方形成巧妙的衔接，并打造了一个开阔、通透的公共空间，为人们的聚会和会友提供了完美空间。大部分的工作区被巧妙地设计成灵活的工作站，可根据建筑的空间布局自由设置。

整个建筑以实现低能耗与良好工作环境之间的完美平衡为设计主旨。基于这一主旨，地板采用了辐射供暖。不同类型的空间温度调节各异，热空气流在几个空间点中得以回收再利用。人工照明设备根据光照强度自动变换色彩。

此外，建筑的第二层创新型表皮采用稀薄的聚四氟乙烯涂层玻璃纤维材料，能够有效进行防风，同时与开阔的窗户形成完美地结合。如此，员工可以自由调节各自的内部小气候，而这对于一个高层建筑空间来说无疑是一个奢侈品。

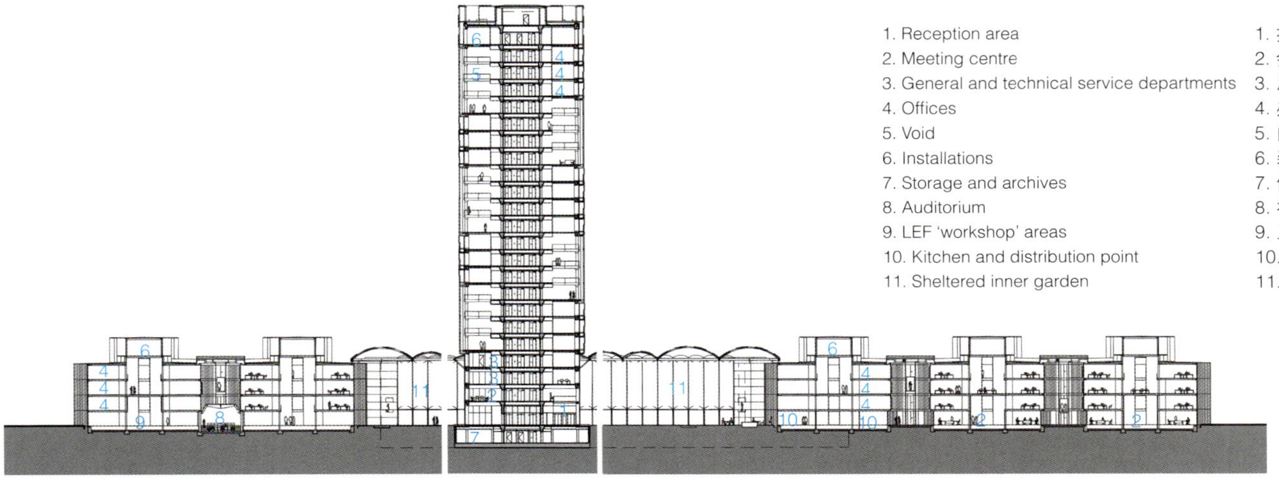

1. Reception area	1. 接待区
2. Meeting centre	2. 会议中心
3. General and technical service departments	3. 总务处与技术服务部门
4. Offices	4. 办公间
5. Void	5. 闲置区
6. Installations	6. 装置
7. Storage and archives	7. 储藏室与档案室
8. Auditorium	8. 礼堂
9. LEF 'workshop' areas	9. 工作间
10. Kitchen and distribution point	10. 厨房和配送中心
11. Sheltered inner garden	11. 隐蔽式室内花园

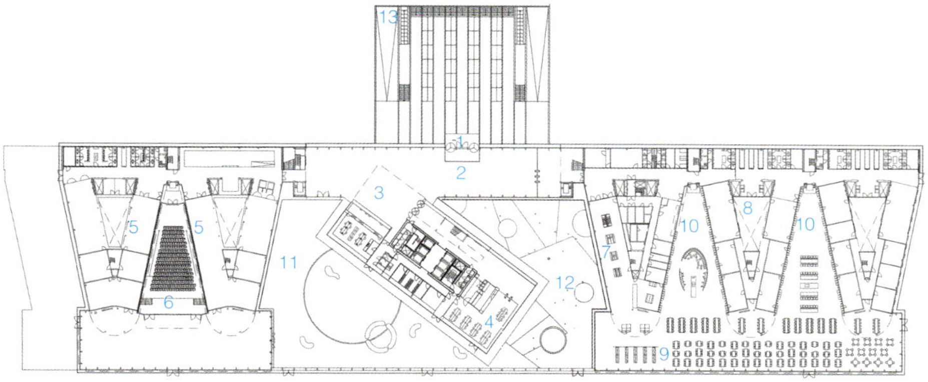

1. Main entrance
2. Entrance passageway
3. Reception desk
4. Reception area, grand café
5. Workshop areas
6. Auditorium accommodation 275 people
7. Distribution point of canteen and kitchens
8. Meeting centre
9. Canteen
10. Foyer of meeting centre
11. White garden
12. Green garden with canteen terrace
13. Entrance to bicycle parking facilities under entrance square

1. 主入口
2. 入口通道
3. 接待台
4. 接待区，大型咖啡厅
5. 工作区
6. 礼堂可容纳275人
7. 餐厅和厨房的配送区
8. 会议中心
9. 餐厅
10. 会议中心休息室
11. 雪园
12. 与食堂平台相连的绿色花园
13. 入口广场下方的自行车停车处入口

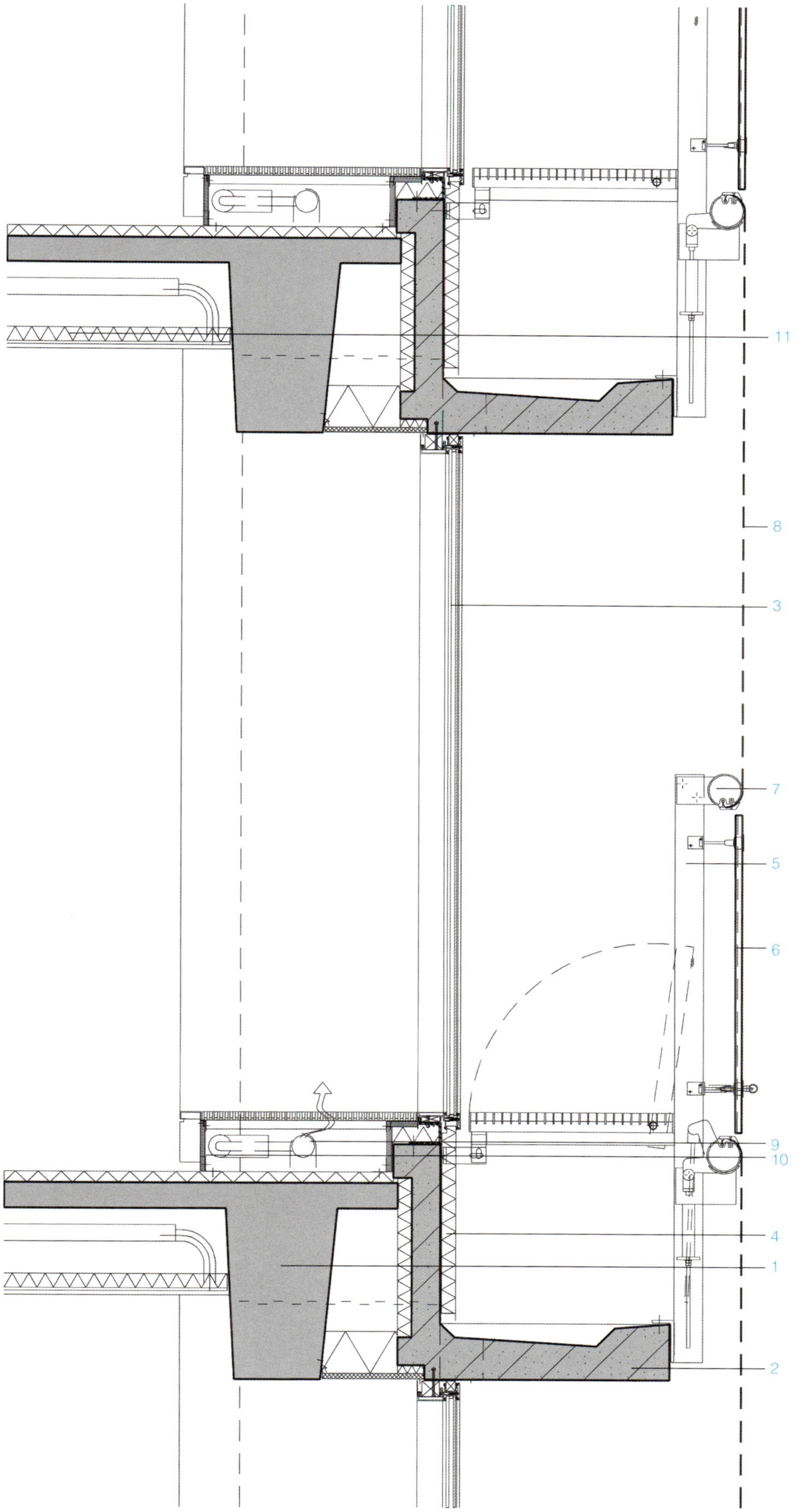

1. Existing concrete floor
2. Existing appended concrete balcony
3. Aluminium frame with clear insulating glass and part that can be opened
4. Sandwich panel to prevent thermal
5. Steel auxiliary construction equipped with anchoring slots for window cleaning carriages
6. Layered and hardened glass
7. Aluminium tubing with twilled profile for attaching and spanning fibreglass fabric
8. PTFE (teflon)-coated fibreglass fabric
9. Heating pipe to prevent cold front
10. Data and electrical connection, supply via column
11. Steel climate ceiling, placed between the concrete beams to ensure the maximum free height; induction units, lighting, sprinklers and loudspeakers have been integrated in the ceiling

1. 原有的混凝土地面
2. 原有的附加混凝土阳台
3. 铝合金窗框中设置了透明中空玻璃，部分可开启
4. 隔热夹层板
5. 配有锚固槽的钢配套结构，便于窗户的清洁
6. 分层和硬化玻璃
7. 带有斜纹的铝材管道系统依附并横跨玻璃纤维构造
8. 聚四氟乙烯（特氟龙）涂层玻璃纤维构造
9. 加热管，能够有效防止寒流的侵袭
10. 数据和电气连接，通过梁柱进行供给
11. 钢架天花板，位于混凝土横梁之间，以实现高度的最大化；感应 装置、照明设备和洒水装置以及扬声器被巧妙地隐藏在天花板之中

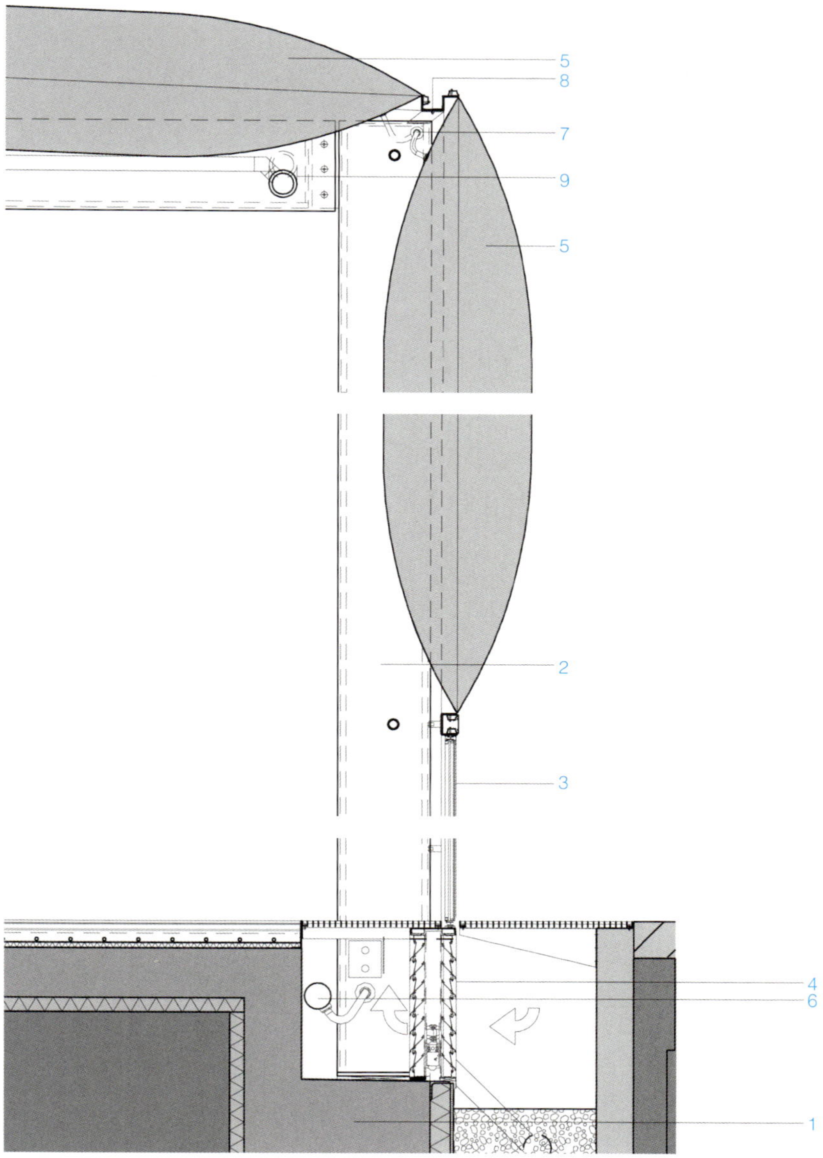

1. Concrete Foundation
2. Steel box column, 8x50 cm, also air supply for the ETFE cushions
3. Steel frame with clear isolatieglas
4. Aluminium ventilation grille
5. ETFE cushions
6. Main pipeline ETFE cushions
7. Steel tube column with air hose
8. Structural steel gutter
9. Sprinkler pipes

1. 混凝土地基
2. 钢箱型柱，8x50厘米，同时也作为四氟乙烯衬垫的空气供给来源
3. 带有透明绝缘玻璃的钢架
4. 铝制通风格栅
5. 四氟乙烯衬垫
6. 主管道的四氟乙烯衬垫
7. 带通风软管的钢柱
8. 结构钢槽
9. 喷灌管道

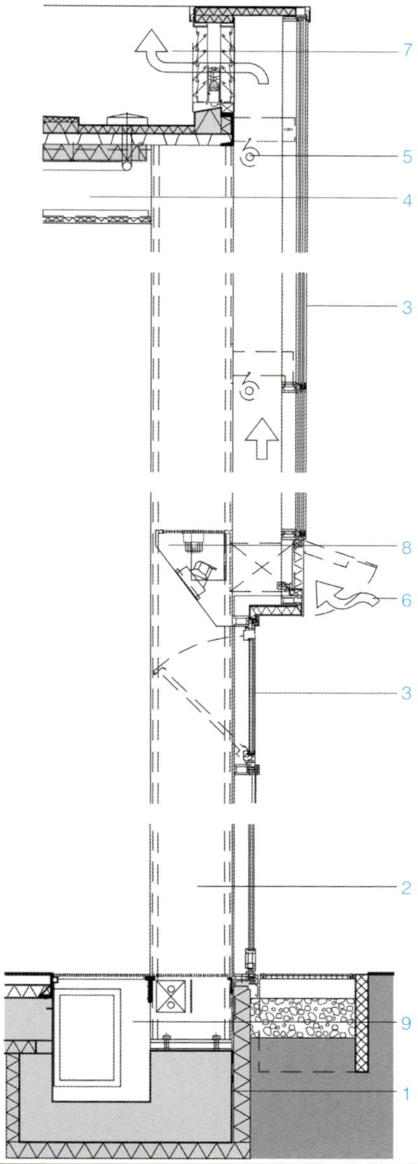

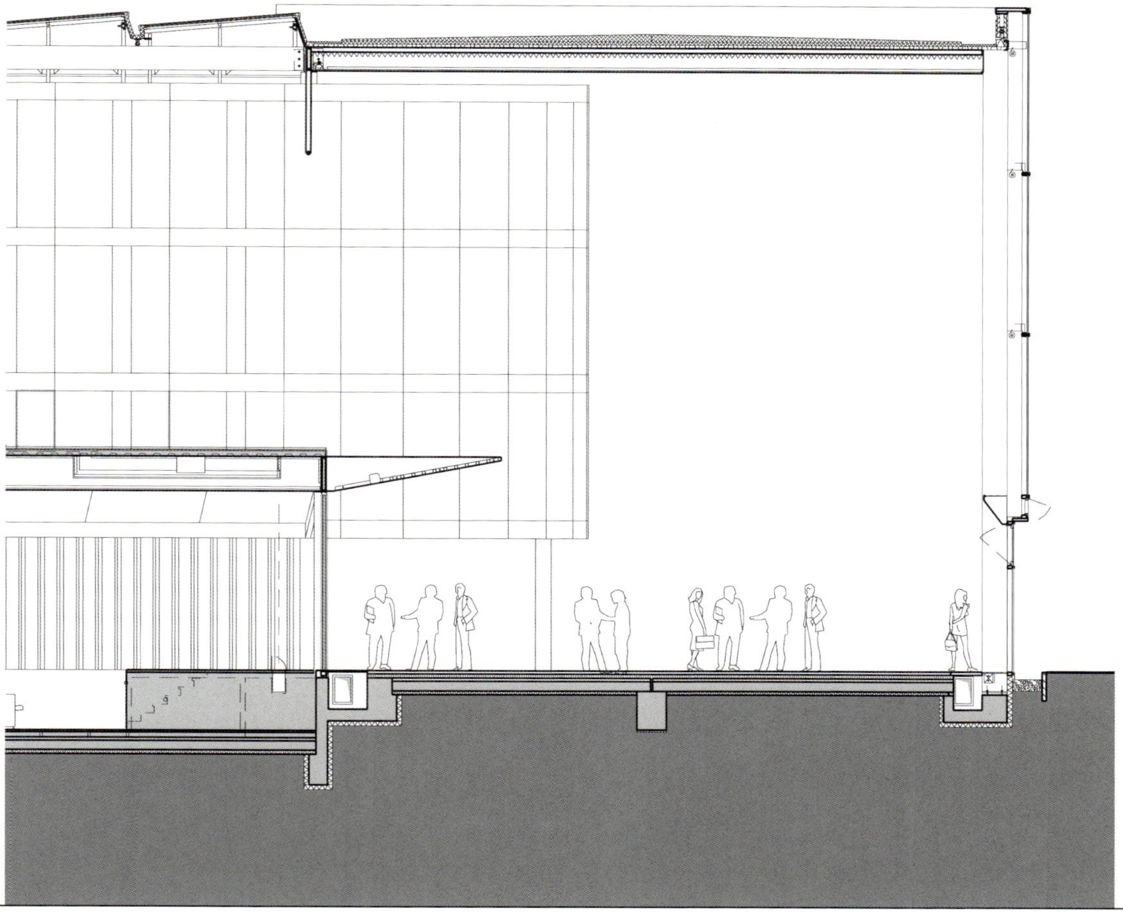

1. Concrete foundation beam
2. Steel box beam 8cmx50cm
3. Aluminium modular façade with insulating and sun-resistant glass
4. Steel roof with a ceiling of perforated, clear-anodised aluminium between the joists
5. Sun-resistant glasshouse fabric with tubular motor
6. Air supply for cavity ventilation
7. Air discharge for cavity ventilation
8. Lightning and convector unit, supply of water and electricity takes place via steel column
9. Convector and main supply track

1. 混凝土地基梁
2. 钢箱梁8x50厘米
3. 模块化铝材外立面上设置了绝缘和遮阳玻璃
4. 钢梁屋顶结构，打孔式天花板，托梁之间以阳极氧化铝为材料进行衔接
5. 带管状电机的耐高温温室结构
6. 通风口空气供给
7. 通风口气体排放
8. 闪电和换流器，通过钢柱提供水和电的供给
9. 换流器和主要供给轨道

Menara Perak

Menara Perak办公楼

Location: Kuala Lumpur, Malaysia
Designers: GDP Architects Sdn Bhd
Photographers: K.L. Ng Photography / RUPAJIWA Studio
Completion date: 2008

项目地点：马来西亚，吉隆坡
设计师：GDP建筑有限公司
摄影师：K.L. Ng摄影工作室，RUPAJIWA摄影工作室
竣工时间：2008年

This development was conceived as an up-market corporate office within the vicinity of the Kuala Lumpur City Centre. As a high-rise office building, the paramount concern for the design team is to achieve efficiency in floor area and energy consumption apart from the creation of a distinctive urban presence at the strategic corner site.
The 26-storey building was designed as a simple triangular box in response to the site profile and constraint. The building comprises a basement floor housing M&E services and car park, ground floor entrance lobby and gallery, 6-storey aboveground car park, 17-storey office and a roof garden on 24th floor. The structural system adopted is conventional beam slab system at the core and post-tensioned beam and slab at the office floor plate to maximise office's ceiling height.
Situated on a very tight corner site, the floor plate was designed to optimise the usable floor area and views towards the city centre. With the service core neatly nestled at the rear northwest corner, the office space is left with maximum glazed wall exposure with minimal intrusion form the columns. The floor plate was also designed to accommodate different tenancy profile, resulting in a flexible, functional and conducive working environment for the tenants.
The building external finishes is a combination of stone cladding on the ground floor entrance, curtain wall at the south and west façades, aluminium fins and louvres form part of the less transparent western and northern façades. The main façade facing south is expressed as a strong curve addressing the approach in curtain wall glazing from level one to the roof terrace. Low-E glazing was used for office floors to reduce heat load and reduce energy consumption whilst polka dots ceramic frit glass was used for aboveground car parks. Façade lighting was also integrated into the curtain wall façade design to enhance salient features of the building. The glowing light boxes on the front façade, in particular are reminiscent of the stainless steel fins, which sparkle in the sun during daytime.

这座奢华的办公大楼位于吉隆坡城市中心附近。作为一个高层办公楼，其内部每个楼层空间的有效利用、能源的低消耗以及特殊地理位置上的独特外观均是设计师关注的重点。
建筑共26层，简约的三角形盒子状结构与周围环境和所在地城市规划要求相得益彰。建筑包括一个容纳M&E服务中心和停车场的地下层、地上一层的入口大厅和走廊、地上6层高停车场、17层办公空间以及一个位于24层的屋顶花园。建筑的结构采用了传统的核心梁板构造，办公区的楼板采用后张拉梁板结构，最大限度地提高办公室的天花板高度。
因建筑坐落于一个非常拥挤的街角，独特的楼板设计能够优化楼层实际使用面积，并确保室内空间将城市中心的美景尽收眼底。服务中心轻盈地坐落于建筑西北角的后身，办公空间外部则大面积地采用了玻璃墙结构，圆柱形外观同时为打造宁静的内部空间和观赏视角提供了有利条件。此外，独特的楼层设计也能够满足空间租赁的需要，为租赁用户营造一个灵活的功能型办公环境。

建筑一楼入口处表面采用石材装饰，南侧和西侧立面以玻璃幕墙为主体，而透明度较低的西部和北部地区则采用了铝片和百叶窗材料。南向的外立面呈优雅的弧线型外观，从一层至屋顶露台全部采用通透的玻璃幕墙结构。用于办公楼层的低辐射玻璃能够有效地降低热负荷，减少能源消耗，而地上停车场则采用了圆点图案陶瓷玻璃材料。外观照明设备与外观玻璃幕墙完美地结合，有力地突出了建筑的设计特色。立面前端闪烁的灯箱与在阳光下熠熠生辉的不锈钢片交相辉映，为空间增添了无限活力气息。

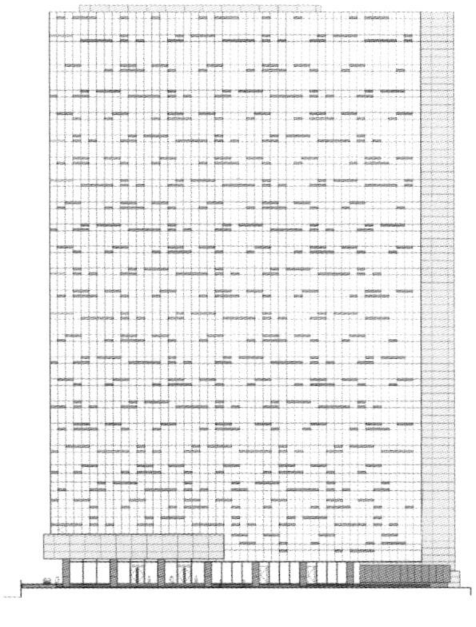

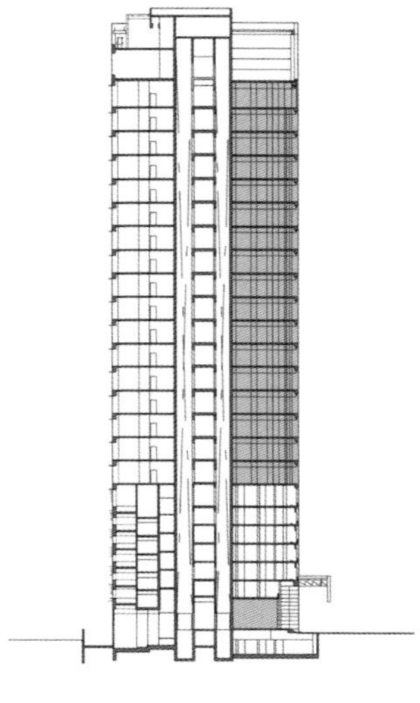

KENANGA

PELBAGAI
PILIHAN
UNTUK ANDA
BANK ISLAM
03-26 900 900

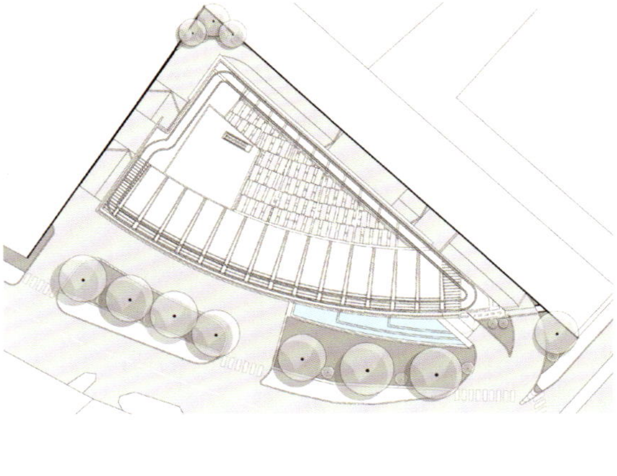

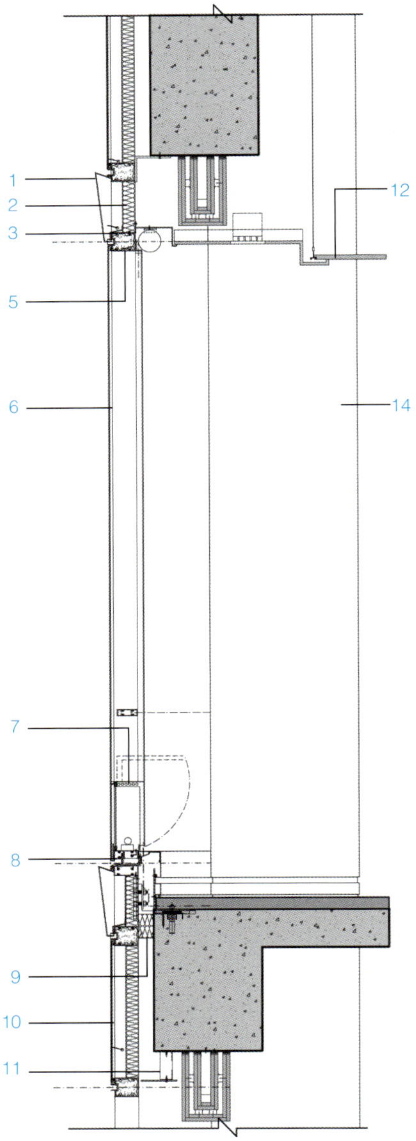

1. Stainless steel reflector
2. 12mm thickness Backpan sealed to frame all round
3. Aluminium plametin powder coat finish
4. Fiber cement board in paint finish
5. Extruded aluminium transom and mullion in powder coat finish
6. 8.76mm thickness laminated glass with hard coat low-E on surface
7. Extruded aluminium sill cover in powder coat finish
8. Structural sealant
9. 2-hour fire rated fire safing insulation
10. Green tinted heat strengthened glass
11. 50mm-thickness Rockwool insulation
12. Suspended ceiling tiles
13. Finish column face
14. RC column face
15. RC crash barrier

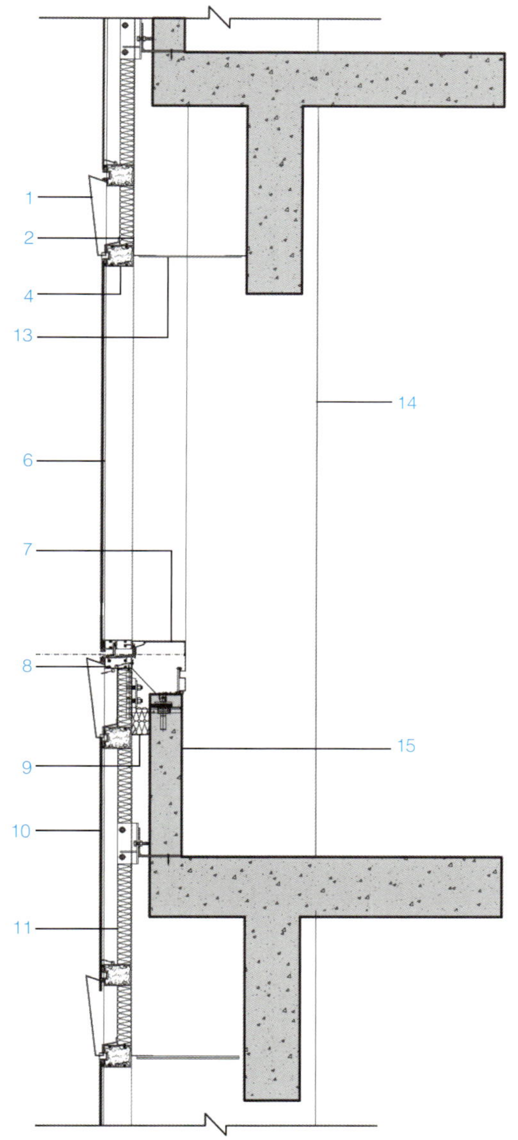

1. 不锈钢反射镜
2. 12毫米厚的水板全面密封框架
3. 铝材粉末涂层
4. 油漆罩面的纤维水泥板
5. 挤压铝横梁和粉末涂层竖框
6. 8.76毫米夹层安全玻璃上设有硬涂层低辐射表面
7. 挤压铝基材粉末涂层表面
8. 密封结构
9. 2小时防火隔热带
10. 绿茶色隔热强化玻璃
11. 50毫米厚的岩棉保温层
12. 吊顶瓷砖
13. 抛光廊柱表面
14. 钢筋混凝土廊柱表面
15. 钢筋混凝土防撞护栏

Palestra

Palestra办公大楼

Location: London, UK
Designers: Alsop
Photographer: Christian Richters
Completion date: 2006

项目地点：英国，伦敦
设计师：艾尔索普建筑师事务所
摄影师：克里斯汀・里赫特斯
竣工时间：2006年

Award name:
2007 RIBA Architecture Award

获奖名称：
2007年英国皇家建筑师学会建筑奖

Palestra offers a very particular response to the speculative office. It offers simple traditional office space wrapped around a central core with floor-to-ceiling glazing and great views. It allows for multi-occupancy, but has in fact been taken by a single tenant - a sign of its successfully simple response to the standard brief. It is a marker building that jauntily pokes its head forward to claim glimpses of the river. Through sheer physical presence and scale it brings an unpromising site to the attention of all, making it the subject of debate.
Palestra is located on Blackfriars Road at the junction with Union Street, opposite the new Jubilee Line extension of Southwark Underground Station. It is at the intersection of four Transport Development Zones.
The brief for the project was to develop a new landmark office building that would provide high quality contemporary office accommodation with flexible floor plates capable of subdivision within a new landmark office building that would not only attract high profile tenants but also make a significant urban contribution to this area of Southwark.
The detailed specification called for standards appropriate for the central London commercial market, specifically a category a fit out, which included air conditioning, raised floors, suspended ceilings, flexibility of plan and rationalisation of stair and lift cores to obtain maximum open floor area.
In addition the client specified the need to incorporate retail opportunities for the ground level of the building. To work successfully in this location this needed:
• An on street presence for A1 retail use to enliven the Blackfriars Road façade
• A contemporary “pod” design to the shop unit with integrated glazing and lighting
• Flexibility in unit size and divisibility
• Separate entrances and services arrangements to the retail unit.

Palestra办公大楼为办公建筑类项目树立了一个别具一格的设计典范。传统的办公空间结构围绕一个中央核心区进行布局，大型落地式玻璃窗巧妙打造出通透的空间效果，并成功淡化了室内外界限。空间运用多功能型设计理念，而实际上内部租赁用户较为单一，而这正与其简约的设计标准相得益彰。Palestra办公大楼是泰晤士河岸地区的地标型建筑。庞大的体量搭配简约、干练的线条轻松吸引人们的眼球，成为该地区的亮点。
Palestra办公大楼坐落于布利费亚斯路与联合大街的交汇处，与萨瑟克区地铁站银禧延长线相对，位于四个交通规划区的枢纽地带。
该项目的设计主旨意在构建一个崭新的地标型办公大楼，高度灵活的波纹板设计能够适应宽敞的空间布局或格状的室内间隔，为用户提供一个充满现代感的高品质办公空间，同时拉动整个萨瑟克区的城市发展。

设计的细节依据伦敦市中心商业市场的规范标准而展开，涉及空调、活动地板、吊顶等，设计灵活、合理的楼梯及电梯槽令空间格外通透、宽敞。
除此之外，一楼的空间中还设置了商业零售区，其成功运营需归功于如下几点：
1. A区的零售店面为朝向布利费亚斯路的立面注入了无限活力。
2. 充满现代气息的“豆荚”式店铺单元区设计与流光溢彩的照明设备完美结合在一起。
3. 店铺单元区的尺寸和分类较为灵活。
4. 零售区拥有独立的入口和服务区。

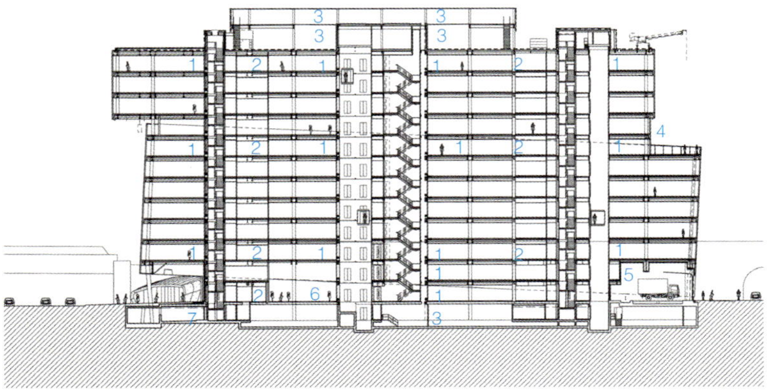

1. Office
2. WC
3. Plant
4. Terrace
5. Loading bay
6. Entrance hall
7. Car park

1. 办公间
2. 卫生间
3. 车间
4. 露台
5. 进货间
6. 入口大厅
7. 停车场

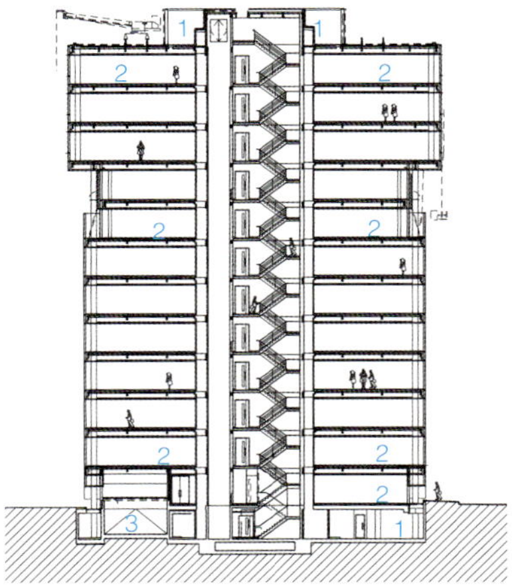

1. Plant 1. 车间
2. Office 2. 办公间
3. Ramp 3. 坡道

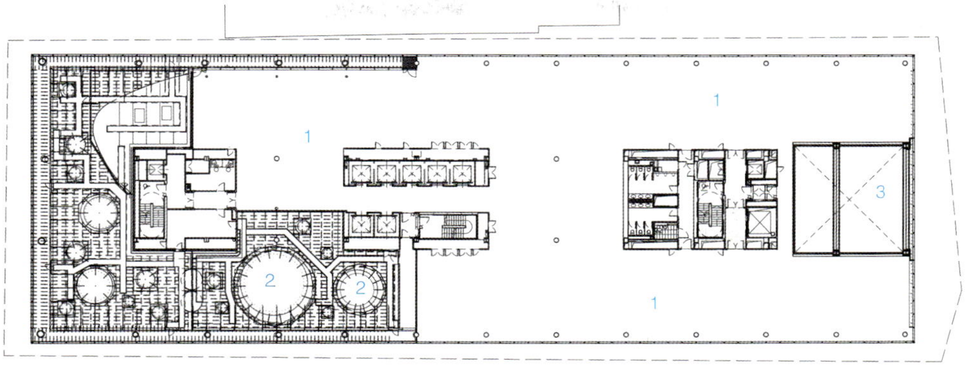

1. Office 1. 办公间
2. Cone 2. 圆台体
3. Void 3. 闲置空间

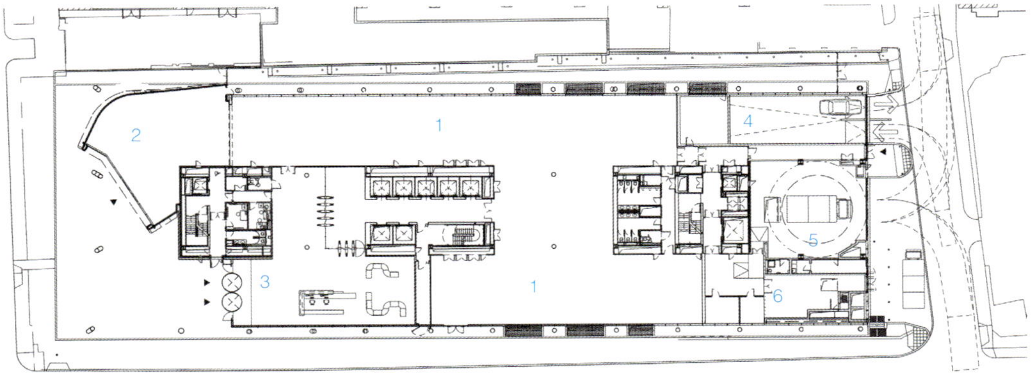

1. Office 1. 办公间
2. Pond 2. 水池
3. Entrance hall 3. 入口大厅
4. Ramp 4. 坡道
5. Loading bay 5. 进货间
6. Refuse 6. 废物区

197
Blackfriars
Road

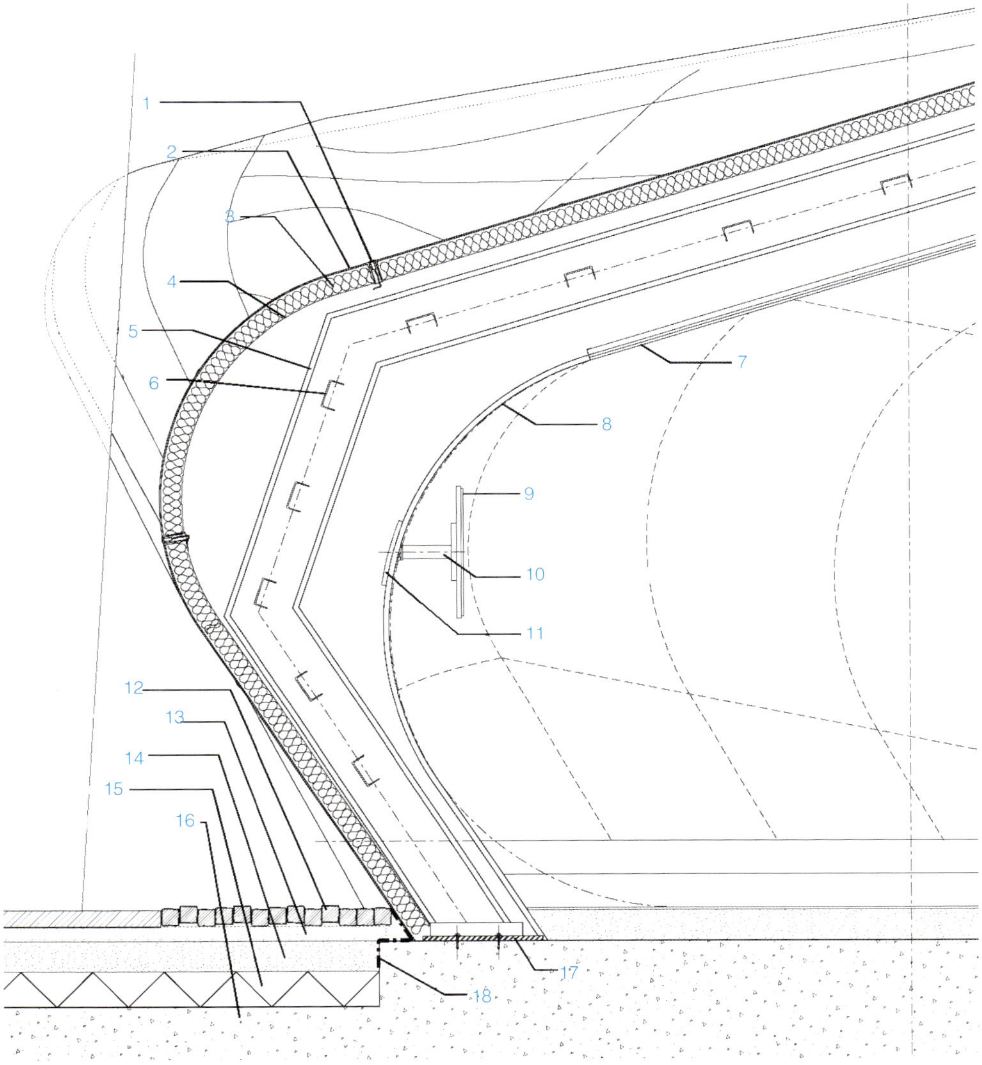

1. LED light fitting
2. 6mm GRP external cladding
3. 75mm renotherm sprayed thermal insulation
4. Line of GRP flange
5. Primary steel secondary steel
6. Secondary steel
7. Double layer plasterboard skimmed and painted
8. 25mm single layer GRG
9. Display panel mounted onto timber board
10. Metal bracket fixing
11. Plywood batten support for fixing
12. Granite setts
13. Sand/cement bed
14. 150mm screed
15. Void former
16. Waterproof concrete
17. Steel plate bolted to concrete
18. GRP dressing

1. LED照明设备
2. 6毫米玻璃钢外部覆层
3. 75毫米喷涂保温绝缘层
4. 玻璃钢边缘
5. 一级钢，二级钢
6. 二级钢
7. 双层脱脂和着色石膏板
8. 25毫米单层玻纤石膏
9. 显示板安装在木板之上
10. 固定金属支架
11. 用于固定的胶合板板条
12. 花岗岩长方砌石
13. 砂/水泥床
14. 150毫米砂浆底层
15. 原有的空地
16. 防水混凝土
17. 钢板混凝土组合
18. 玻璃钢表面修光

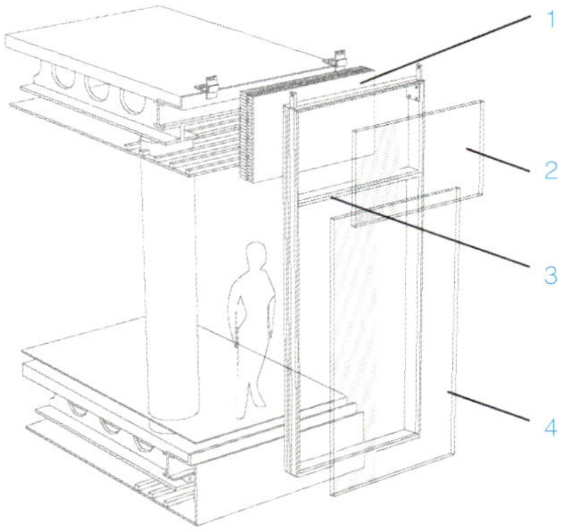

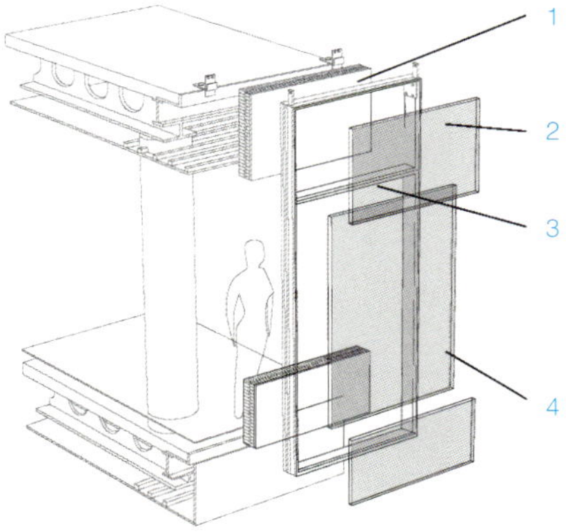

1. Insulated spandrel panel with ventilated shadow gap. Panel finish varies. Refer to architectural elevations
2. Spandrel double glazed unit with toughenec glass
3. Intermediate transom, double glazed units are finished externally with a butt silicon joint
4. Main vision double glazed unit with toughened outer and laminate inner

1. 带有通风遮蔽空隙的绝缘拱肩楼板，建筑立面板材抛光程度各异
2. 带有钢化玻璃的双层拱肩玻璃单元
3. 中间横梁和双层玻璃窗单元的外部通过一个对接硅结合点连接在一起
4. 主要透光双层玻璃单元外部采用不透明处理，内部采用锻压处理

Mediacomplex

Mediacomplex办公大楼

Location: Barcelona, Spain
Designer: Patrick Genard
Photographer: Aleix Bagué
Completion date: 2008

项目地点：西班牙，巴塞罗那
设计师：帕特里克・戈纳德
摄影师：艾蕾丝・博格
竣工时间：2008年

Award name:
World Architecture Festival 2008 Selected Project
Premios Ciudad de Barcelona 2008, etc.

获奖名称：
2008年世界建筑节入围
2008年巴塞罗那城市奖等

The Mediacomplex project was born from the union between a city, the 22@ Plan for the development of emergent technologies, and a company as engine within this sector.
The project addresses a complex programme which integrates a variety of diverse elements. These include many facets of the audiovisual world, the teaching of the University Pompeu Fabra, the production within the Factory of images, and the creation within the Tower, Itself a display window upon the Ave. Diagonal. Each of these elements occupies its place within the building autonomously, and still participates with the same whole.
The building occupies a special place. The meeting point between the orthogonal grid of the Ensanche and the Ave. Diagonal creates a new New York-style pedestrian square, generously opening up the audiovisual building to the city. In addition, the project positions itself over the Ave. Diagonal as an entryway into the "peri Glories" zone which finds its counterpoint with the Agbar tour. Lastly, it takes advantage of the industrial history of the neighbourhood by incorporating the unfinished 19th century factory of Cal Aranyo.
From this integrated programme, keeping in mind the urban complexity of the site, two new buildings are born in synergy. Their polar characteristics enter a dialogue allowing the whole to become greater than the sum of its parts.
A base, the Factory, static and horizontally treated, is aligned with the old existing Cal Aranyo factory as a natural continuation two centuries later. The Tower, vertically treated, dynamically expresses the change between urban alignments. The base, opaque and smooth, covered with a double-skinned metallic screen, provides subtle variations of permeability and perforations which softly filter the light. The Tower, transparent, expresses its deeply recessed structure with huge niches, creating chiaroscuro effects on the façade. Nevertheless, both share a common system of geometry and proportion, as with a metallic and chromatic treatment that achieves a unity of rich diversity.

Mediacomplex办公大楼是巴塞罗那城、22 @新兴技术发展规划以及该行业的一个先锋企业共同合作的结晶。
该综合型办公大楼中集合多种元素于一体。这些元素涉及音像制品、庞培法布拉大学的教义、影像制作工厂产品、街面上大楼橱窗的陈列品。这些元素在各自的领地中扮演着独特的角色，并相互融合，打造了一个和谐的统一体。
该建筑所在地地理环境较为特殊，其位于Ensanche广场区和高速铁路之间的垂直交叉线上。对角线地区巧妙打造了一个崭新的纽约风格步行广场，将视听大楼向整个城市敞开。与此同时，建筑凌驾于高速铁路之上。对角线地区作为“辉煌边界区”的入口与阿格巴塔大楼遥相呼应。
该综合型建筑充分结合所在地复杂的城市环境，两幢建筑和谐互融，鲜明的特色对比完美地实现了空间的对话，打造出的整体美感远超过各部分之和。
位于建筑下方的工厂沿水平方向布局，彰显静态之美，并与两个世纪前的Cal Aranyo老工厂自然衔接。塔楼呈垂直状设计，动感十足，完美表达出城市道路定线之间的变化。工厂大楼的外观光滑而不透明，外部覆以双层金属遮屏，确保室内通风、光线充足。相比之下，塔楼的外观较为通透，嵌壁式结构形成的大型壁龛打造出明暗交替变幻的立面效果。然而，相近的几何形状和比例以及巧妙的金属材料和色彩运用恰恰实现了这些多样化元素的统一。

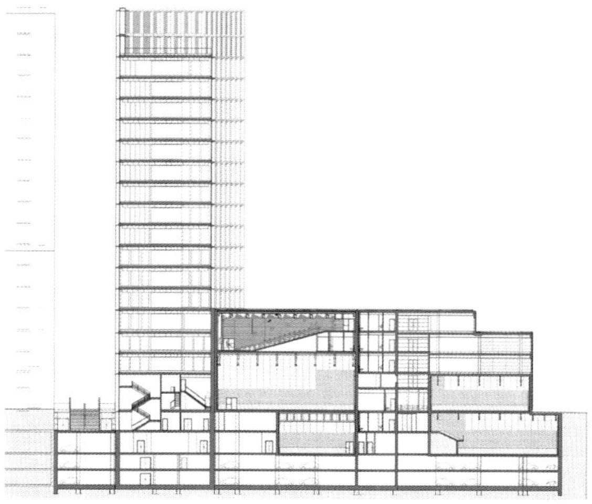

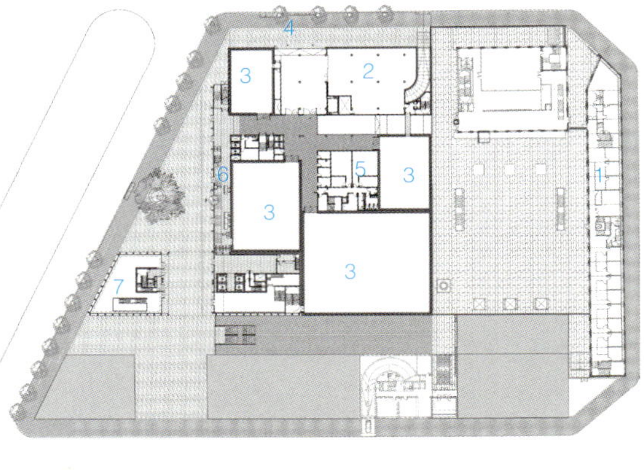

1. Antique building
2. Sets storage
3. TV set
4. Sets entrance
5. Technical control
6. Main access to broadcast centre and tower
7. Office tower with terrace on the 16th floor

1. 老建筑
2. 储藏区
3. 视听区
4. 入口
5. 技术控制区
6. 广播中心和大楼主要入口
7. 带有露台的第16层办公大楼

imagina

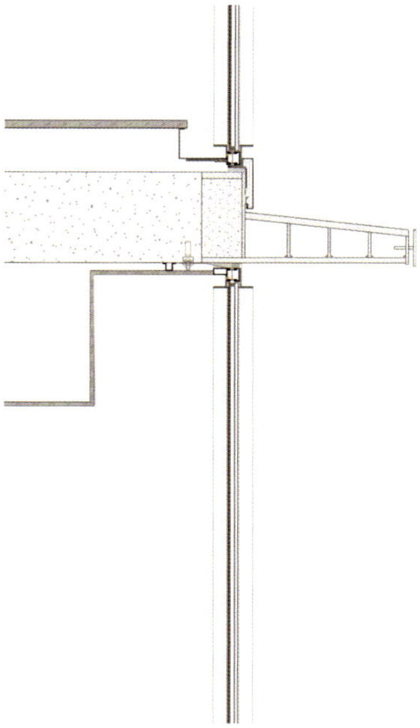

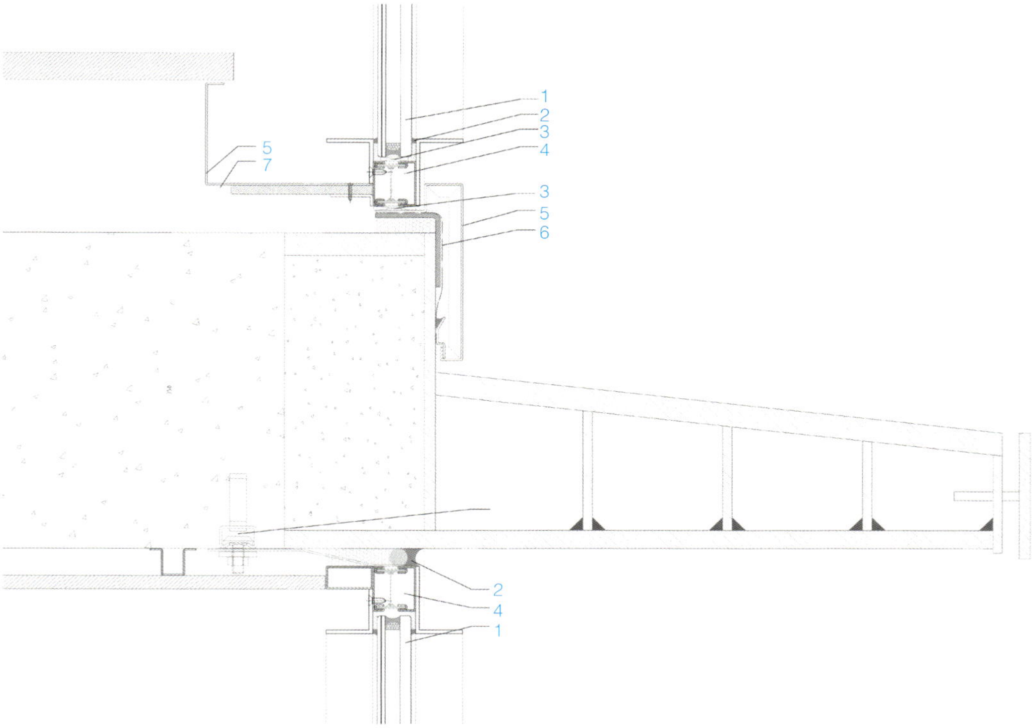

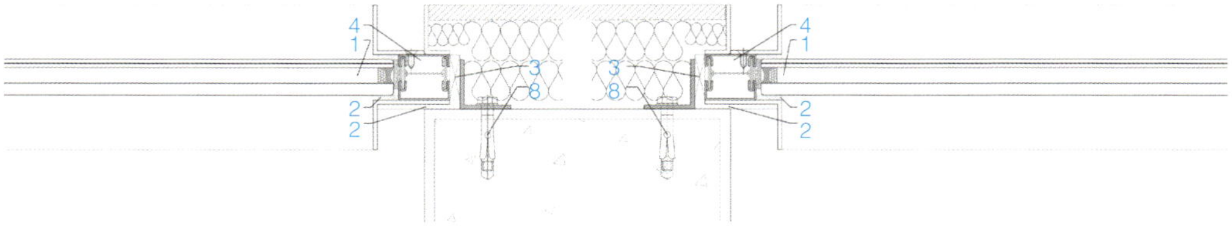

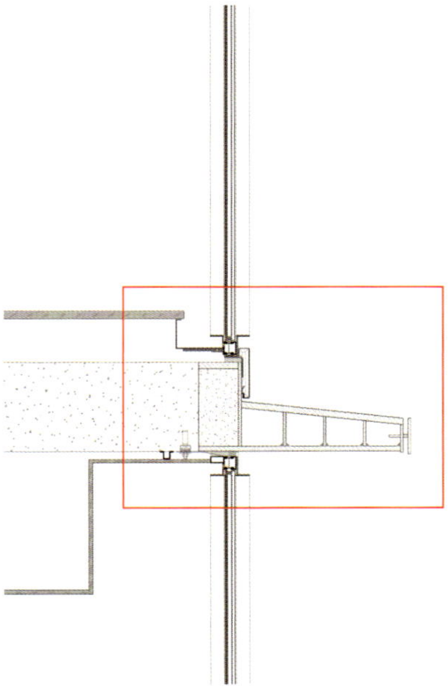

1. Security double glazing low-E solar control
2. Silicone seal type dc-797 neutral colour s/df, with bottom of polyethylene gasket
3. Rubber gasket or shim EPDM
4. Fixed steel window Jansen Janisol special type, including Hammerite Martele interior and exterior ends
5. Trim galvanised steel E=1.5mm, shaped and machined, including Hammerite Martele interior and exterior ends
6. Waterproofing sheet of EPDM rubber E=1mm placed in cold
7. 16mm OSB molding reinforcement in accessible areas
8. Taco metal expansion

1. 安全中空玻璃，低辐射控制
2. 中性DC-797型密封硅胶，颜色s/df，底部为发泡聚乙烯垫片
3. 橡胶垫片或三元乙丙垫片
4. 詹森·哈尼索尔特别型的固定塑钢窗，包括哈梅利特·马尔特雷室内及室外部分
5. 已定型加工的E=1.5毫米装饰镀锌钢板，包括哈梅利特·马尔特雷室内及室外部分
6. 冷却的E=1毫米三元乙丙橡胶防水板
7. 在通行区域铸模加固的16毫米定向刨花板
8. 塔科金属延伸管

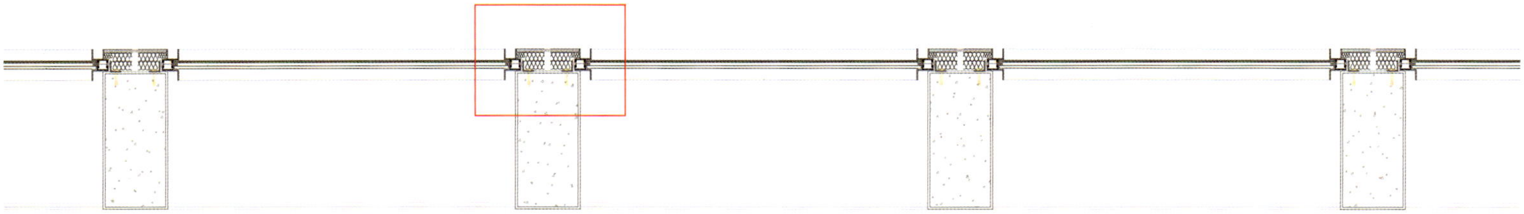

UNIQA Vital Business Centre

优尼卡保险公司重要业务中心

Location: Budapest, Hungary
Designers: Ferdinand and Ferdinand Architects
Photographer: Istvan Oravecz
Completion date: 2009

项目地点：匈牙利，布达佩斯
设计师：费迪南德&费迪南德建筑师事务所
摄影师：伊什特万·奥来维克兹
竣工时间：2009年

The new head office is located close to the previous one at the junction of two important traffic arteries in District XIII, Lehel út and Róbert Károly körút. The office building constructed by property developer Raiffeisen Evolution in 22 months, named the Vital Business Centre, and has net usable floor space of 18,000 square metres. That means it is of a similar size to the Austrian Foreign Ministry. The nine-storey building houses all of UNIQA's divisions, while 40% of the usable floor space will be let to other companies.
Companies renting the space so far include a pharmacy and a diagnostics centre. Árpád Ferdinánd explained that while designing the building he was guided by the principle of transparency. Transparency is one of the cornerstones of UNIQA's business philosophy, and is conveyed in the Vital Business Centre by many glass surfaces and expansive rooms, the architect said. Lighting up the night a distinctive feature of the building is its LED façade, allowing colourful images and messages to be displayed on the street-facing side of the office building every evening. Although the idea is not new, it is unique in its scale: with some 80,000 pixels the Vital Business Centre currently boasts Europe's largest LED façade.
There is a 50 metres long and 20 metres high cube outstanding from the 3rd floor up in the middle of the 95metres façade to break the layout of it. This cube is two floors higher than the building itself so the visual effect is like a box was pushed out of the house. The main entrance has been placed under this cube. The elevation was made of aluminium-glass certain construction. To break the rhythm of the vertical profiles the architects designed random green stone bands. They have had to isolate because of the different chemical behaviour of stone.
Only the middle part of the façade contains the LED lighting system. The panels have been built in under the windows between a concrete wall and the glassing. One panel is 25 centimetres x 25 centimetres; there are six floors with this built-in system.

新总部位于匈牙利布达佩斯第十三区Lehel ú t大街和R ó bert K á roly kör ú t大街的交汇处。该办公建筑由资产开发商Raiffeisen Evolution在22个月内打造完成，命名为优尼卡保险公司重要业务中心，实际使用面积为18,000平方米，这与奥地利外交部的面积接近。九层的空间分布着优尼卡保险公司重要业务中心的各个部门，40%的实用楼面空间留作其他公司所用。
其他的租赁用户目前包括一个药房和一个门诊中心。设计师阿帕德·费迪南德解释说，整个建筑的设计以透明度原则为导向。透明化是优尼卡保险公司的经营哲学之一，通过通透的玻璃界面和开阔的房间设计将这一哲学彰显得淋漓尽致。夜幕降临之时，独特的LED显示屏将外立面打造得色彩斑斓、美轮美奂，每晚在临街方向向行人显示出交替变换的彩色图像和信息。尽管这一设计并不算新颖、独特，其庞大的规模却使其脱颖而出：80,000像素点的运用堪称是欧洲当今最大的LED照明外立面。
建筑外立面的中央从四楼开始出现的一个50米长、20米高的立方体体量打破了整个建筑的布局。该部分比整个建筑高出两层空间，因此，远远望去犹如一个从建筑中抛出的盒子。主入口位于该部分空间的下方。立面采用铝－玻璃材料构成。为打破垂直剖面图呆板的布局节奏，设计师巧妙地打造出自由分布的绿石带，因该石材化学成分的不同而互相分隔开来。
仅置于外立面中央部分的LED照明系统设于窗户下方的混凝土墙和玻璃之间。其中一个嵌板的尺寸为25x25厘米，该系统分布于六层空间之中。

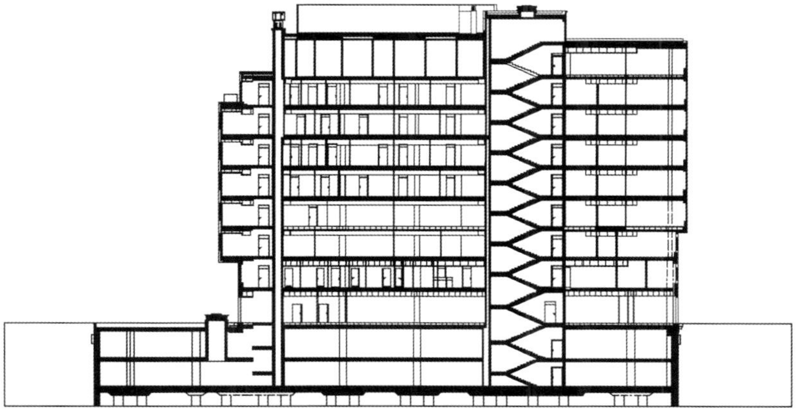

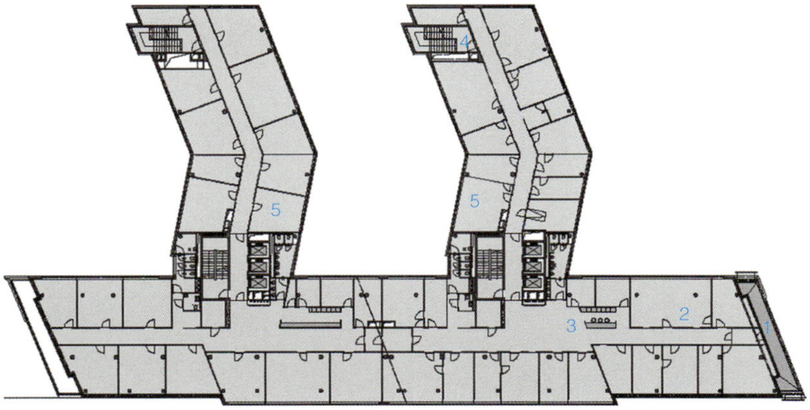

1. Entrance
2. Lounge
3. Reception
4. Management
5. Office area

1. 入口
2. 休息室
3. 接待台
4. 管理处
5. 办公间

UNIQA
UNIQA

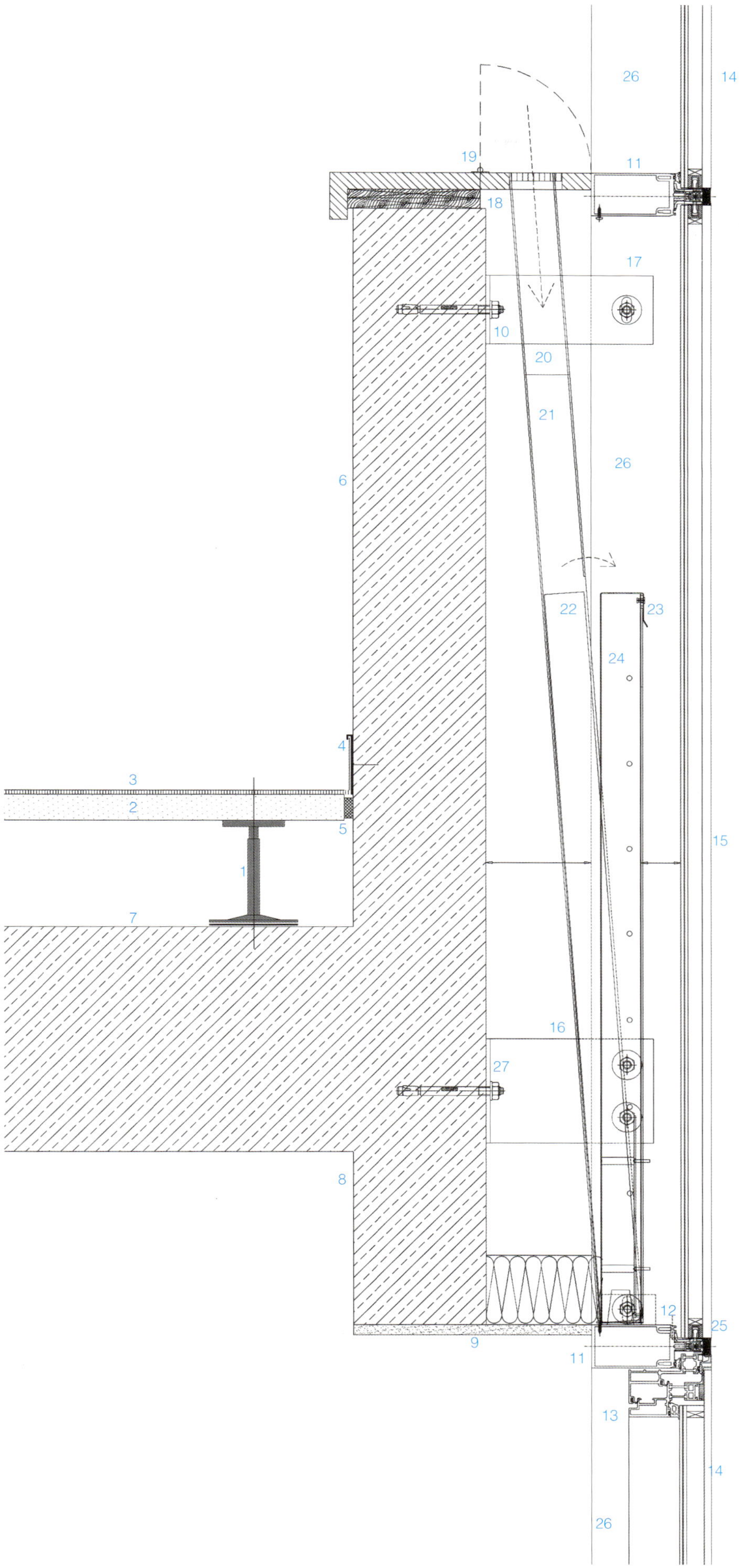

1. Raised floor column
2. Raised floor board
3. Wall-to-wall carpet
4. Skirting
5. Elastic filler
6. Reinforced concrete parapet wall
7. Reinforced concrete slab
8. Reinforced concrete beam
9. Gypsum board 125mm
10. HILTI HST M10x110/30cm
11. Curtain wall venting parapet frame
12. Curtain wall rib fixing element, moving joint
13. Casement frame, propriate for curtain wall system
14. Heat insulated glazing
15. Heat insulated glazing with opal foil by factory
16. Galvanised steel bracket – with fixed joint
17. Galvanised steel bracket – with expansion joint
18. Internal window sill support
19. Internal window sill with perforated, up opening Section
20. Fixing rail of lighting panel in direction of intaking
21. Fixing rail of lighting panel in final situation
22. LED lighting panel in temporary situation
23. Fixing of LED lighting panel
24. LED lighting panel in final situation, fixed
25. Outtaking of moisture
26. Heat insulated aluminium curtain wall rib

1. 活动地板柱
2. 活动地板板材
3. 满铺地毯
4. 踢脚线
5. 弹性填料
6. 钢筋混凝土护墙
7. 钢筋混凝土板
8. 钢筋混凝土梁
9. 125毫米厚石膏板
10. 螺栓式锚栓M10x110/30厘米
11. 幕墙通风矮墙框架
12. 幕墙挡边固定元素，活动结合点
13. 幕墙系统的窗扉框
14. 隔热玻璃
15. 蛋白石隔热玻璃
16. 镀锌钢支架 –固定点
17. 镀锌钢支架 –与伸缩缝
18. 内部窗台板支座
19. 内部窗台板，附有打孔向上开口断面
20. 入口方向的照明板固定围栏
21. 后方的照明板固定围栏
22. 临时地区的LED照明板
23. LED照明板的固定
24. 后方的LED照明板，固定
25. 排气孔
26. 隔热铝幕墙挡边

Creative Valley

创意谷

Location: Utrecht, The Netherlands
Designer: MONK & Paul Dam
Photographer: Trevor Mein
Completion date: 2010

项目地点：荷兰，乌得勒支
设计师：蒙克设计事务所，保罗・达姆设计事务所
摄影师：特雷弗・迈因
竣工时间：2010年

Creative Valley is a building for companies that want to work innovatively, mostly cultural creatives. Gathering their wishes the designers needed to produce a sustainable multi-talented building for multiple companies, each with its own identity yet sharing a certain number of facilities and contacts - an environment in which accommodation and business are merged.

The appeal of Creative Valley is that people feel completely at home from the moment they enter the building and experience an "open energy".

MONK architects created a work environment that people enjoy being in as well as a building which is future-proof. The design is in fact a literal drawing of the proposed organisation chart. The architecture was inspired by the natural structure of a tree with branches and leaves. It has an enclosed midsection with heavy structural elements and materials in monolithic detail. Boxes with an overhang of eleven metres are suspended from the midsection that defy gravity and offer maximum views by means of layered glass façades.

The client wanted to realise a building for creative users containing public, privileged and private space. Public space is primarily meant for meeting purposes; privileged spaces are the meeting rooms and presentation spaces; the private spaces are the individual business areas.

This resulted in an elongated, relatively narrow (5.4 metres) nave with seats, meeting rooms, WIFI and pantries. The heads of the nave are meeting and presentation rooms in various sizes and styles. On either side of the nave the glass extensions are hung in which the individual business units are located. On the roof of the lower extensions are common roof terraces.

创意谷建筑专为创意公司，尤其是文化创意公司而设计。在集思广益之后，设计师提出了一个为多个公司打造可持续性多功能建筑的方案，力求每个空间在共享部分设施并保持互动的条件下，能够彰显公司的品牌独特性，融居住和办公于一体。

创意谷的魅力在于使人们在迈进空间的瞬间即刻感受到如家般的温暖，体验开放、充满活力的空间之感。

这个由蒙克设计事务所塑造的独特环境在为人们提供舒适怡人的氛围的同时，能够经受住时间的考验，历久弥新。实际上，设计是提议的组织结构图的一个书面表达形式。建筑的设计灵感源自对枝叶繁茂的大树的参考。封闭的中央空间中采用厚重的建筑元素和材料，强调空间的整体性。11米高的悬臂从中央部分悬垂下来，顽强地抵抗着地心引力，并通过层次化的玻璃立面开拓空间的视野。

客户要求建筑中能够为创意用户提供公共区域、专门空间和私人空间。公共空间主要作为召开会议之用；专门空间主要为会议室和报告室而提供；而私人空间则为个人办公区域。

根据提议，设计师巧妙地打造了一个狭长的（5.4米）中央广场，其中包括座位区、会议室、无线上网区和茶水间。广场的前端是不同规模和风格的会议室与报告室。广场一端的玻璃延伸区悬垂于个人办公单元区之中。低处延伸区的屋顶设置了共享屋顶露台。

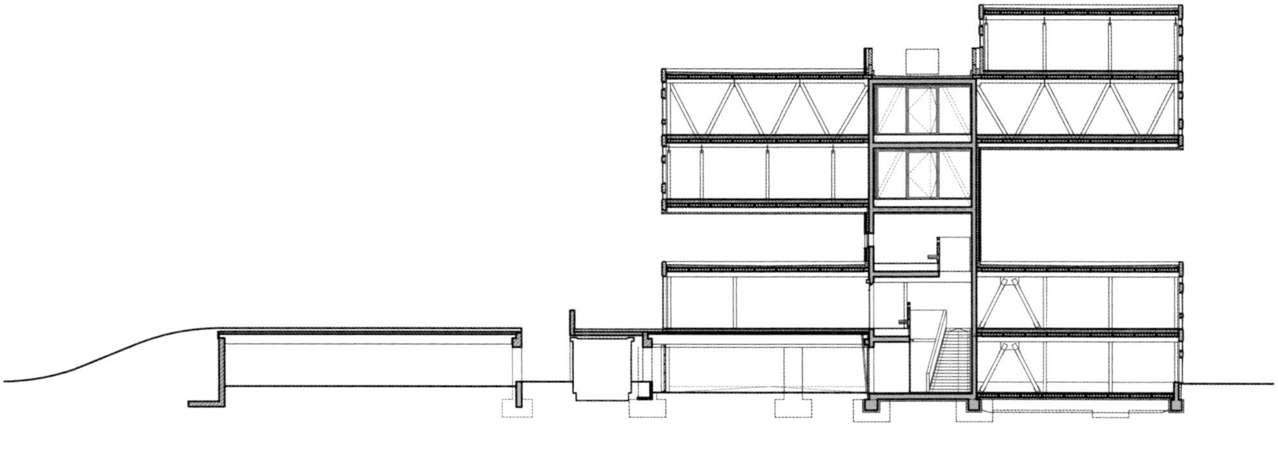

HARTE
WELKOM IN
CREATIVE VALLEY

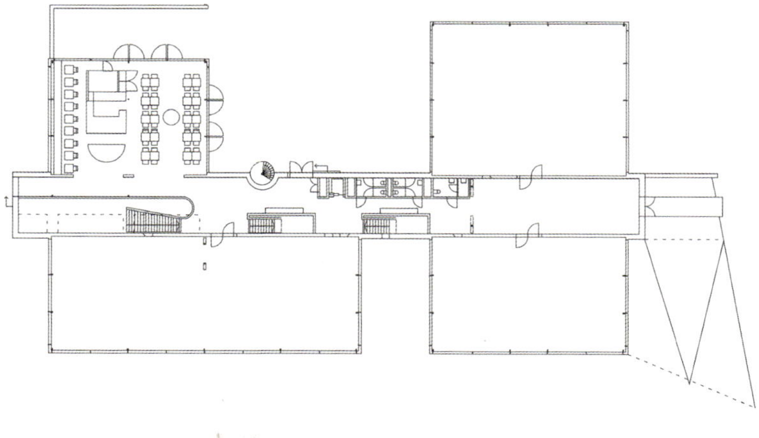

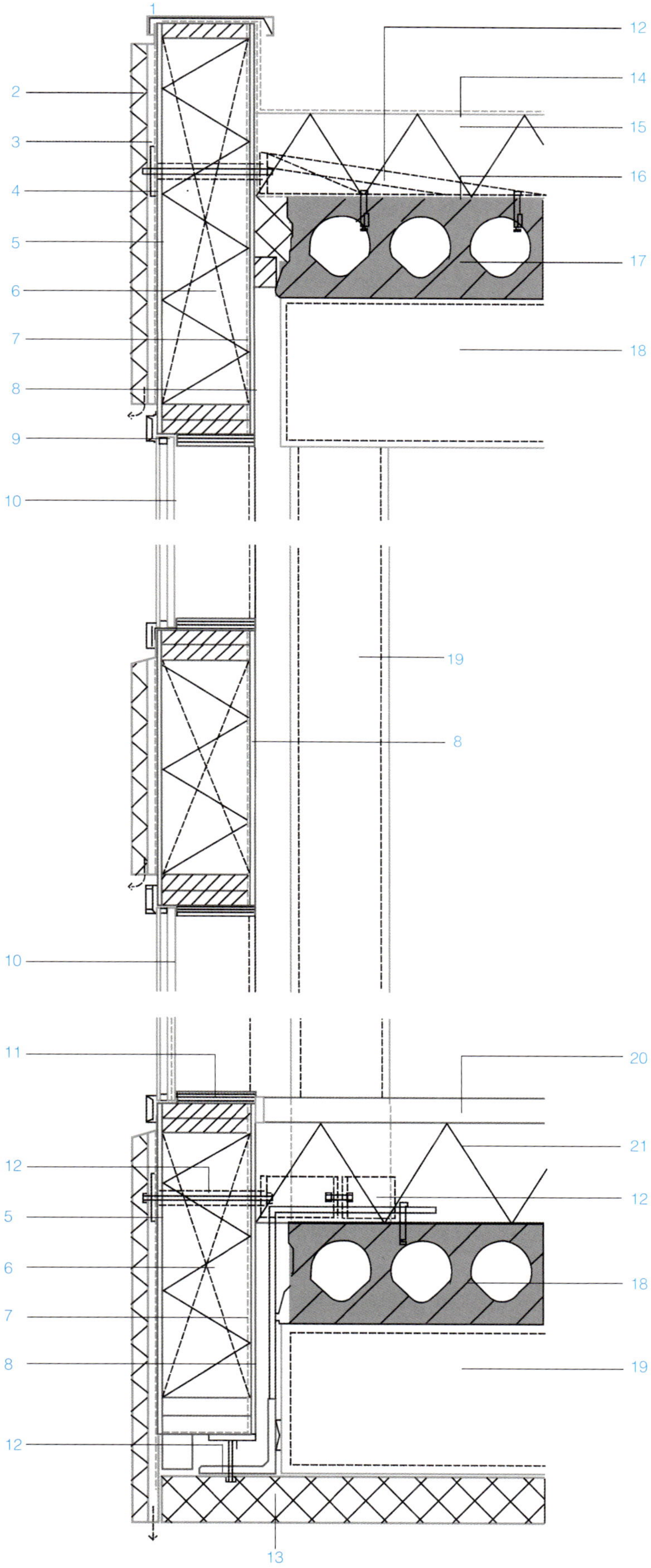

1. Aluminium truss
2. Aluminium sandwich panel isolated 30mm
3. Cavity 20mm
4. Damming building plastic
5. Plywood construction plate 10mm
6. Wooden façade frame and insulant 170mm
7. Damp resisting building plastic
8. Plywood inner plate 12mm
9. Aluminium clamp profile
10. Double-glazed
11. Plywood 15mm
12. Steel anchorage façade frames
13. Aluminium sandwich panel insulated 95mm
14. Asphalt-impregnated paper roofing
15. Insulant
16. Damp resisting layer
17. Precast concrete floor200mm
18. Steel sleeper hollow tubular profile 300mm
19. Steel round column section 193mm
20. Cement screed 50mm
21. Insulation eps 200mm

1. 铝合金桁架
2. 独立的30毫米铝夹芯板
3. 型腔20毫米
4. 建筑塑料隔墙
5. 10毫米胶合板建筑施工板
6. 木制框架立面和170毫米绝缘层
7. 耐湿热建筑塑料
8. 12毫米的胶合板内板
9. 铝合金压板
10. 双层玻璃
11. 15毫米胶合板
12. 钢锚立面框架
13. 95毫米铝合金隔热夹芯板
14. 沥青绝缘浸渍纸屋面
15. 绝缘材料
16. 防潮隔水层
17. 200毫米预制混凝土底板
18. 300毫米钢枕空心管型材
19. 193毫米圆钢柱截面
20. 50毫米水泥沙浆底层
21. 200毫米可发性聚苯乙烯绝缘层

Basque Health Department Headquarters

巴斯克卫生署总部

Location: Bilbao, Spain
Designer: Coll-Barreu Arquitectos / Juan Coll-Barreu / Daniel Gutiérrez Zarza
Photographer: Aleix Bagué
Completion date: 2008

项目地点：西班牙，毕尔巴鄂
设计师：Coll–Barreu建筑师事务所，胡安·科尔·巴鲁，丹尼尔·古铁雷斯·萨尔萨
摄影师：艾利斯·巴格
竣工时间：2008年

Award name:
Nomination to the European Union Prize for Contemporary Architecture - Mies van der Rohe Award 2009
Selected in the X Biennial of Spanish Architecture and Urbanism Basque 2009

获奖名称：
2009年当代建筑欧洲联盟奖——密斯凡德罗奖提名
2009年第十届西班牙巴斯克建筑与城市化设计双年展入围

The site is located in the crossroad of two important streets of the Ensanche, designed in 1862. The restrictive city rules compel to repeat the shape of the neighbouring walls, reducing penthouses according to a curved directive, chamfering the corner and building a tower on it.
The building concentrates services and communications in a vertical spine attached to the longest party hedge and generates seven open floors assigned for offices. Above this, there are two floors for local representative and institutional use. The board hall takes up the double height of the tower. The assembly hall, its lobby and its appendages are situated in the first basement. Further below there are two parking floors and one fourth level for archives. The car lifts allow access to all the basement levels.
The double façade solves not only urban requirements but also those concerning energetic, fire-resistant and acoustic insulation from outside. This climatic improvement enables the elimination of the conventional air-conditioning installation as well as the false ceiling. Thus, the sound produced by the building is reduced, air recirculation in workplaces disappears, with a significant increase of health conditions. The volume occupied per floor is also reduced.
The folded façade generates multiple visual from inside to the streets below, and also from the highest floors to the landscape that surrounds the city, a highly effective mechanism for the incorporation of urban vitalisation inside the building. The workspace benefits of the permeable, passable and livable volume of the façade, that enables the building breathing and the space exchange between inside and outside. The system façade of the building is similar to the experience of sitting at the door of a house, above the threshold, with an eye toward the road and the back into the home.

巴斯克卫生署新总部位于毕尔巴鄂扩建区（成立于1862年）上最重要的两条大街的十字路口。现有的城市规划限制原则要求新建筑能够于周围的墙体形成和谐的统一体，同时，减少阁楼、倒角弯道和上升塔的设置，打造曲线形的建筑外观。
建筑将服务区和交流区集中于一个垂直的“脊柱”空间之内，该部分空间附属于最长的聚会篱栅区，最终生成了七层开放式办公空间。办公空间的上方两层空间专供机构和代表处使用。董事会大堂纵向贯穿建筑的始终。礼堂的大厅和附属空间位于地下室的一层之中。礼堂的下方设置了两层停车场和一个档案室。汽车电梯与地下室所有的楼层相通。
双立面的设计不仅满足了区域规划的要求，同时也解决了节能、防火和隔声的问题。气候的自我调节功能摆脱了传统的空调设施的安置以及假吊顶的设计。因此，健康舒适的办公环境有效降低了声音的分贝，并避免了工作间空气的重复流通。从而，每个楼层所占用的体量面积也大大减少。
折叠的建筑外观在空间的内部及其下方的街道中打造了多样化的视觉效果，并将高层空间与环绕城市的景观有机衔接在一起，在建筑的内部形成了一个高效的城市化机制。通透、轻盈、活力四射的外立面为办公环境带来了积极的影响，有效地加强室内空间的流通，并拉近室内外空间的距离。遥望建筑，其立面系统犹如一个人坐在一间房子的门槛之上，面向大街、背向房间，尽情享受着生活的闲适。

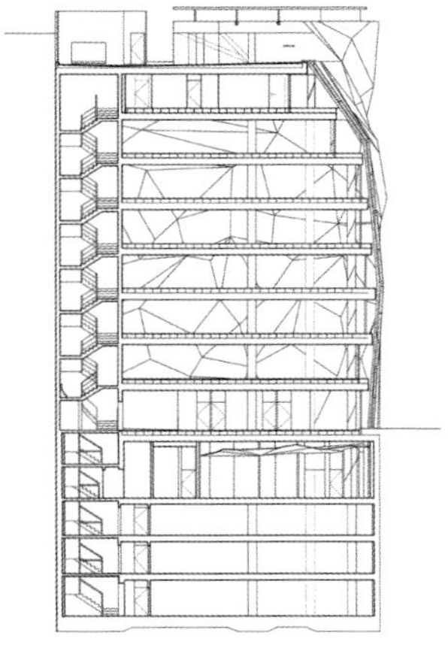

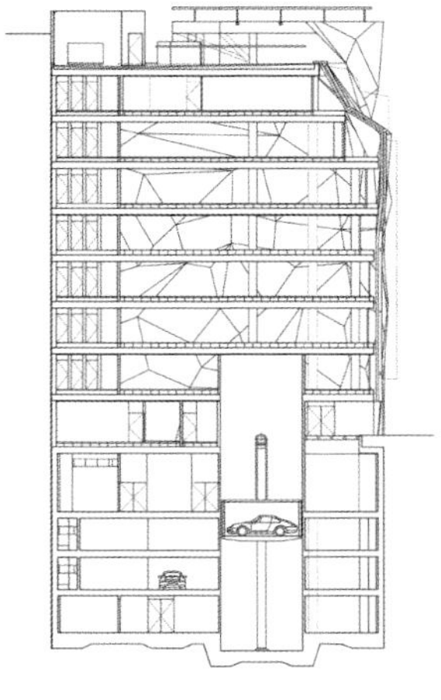

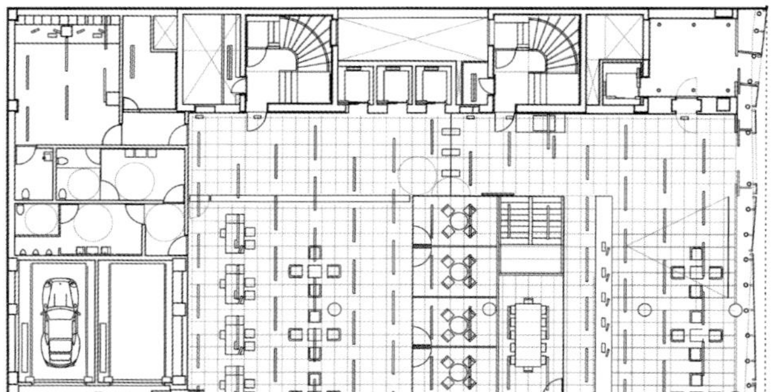

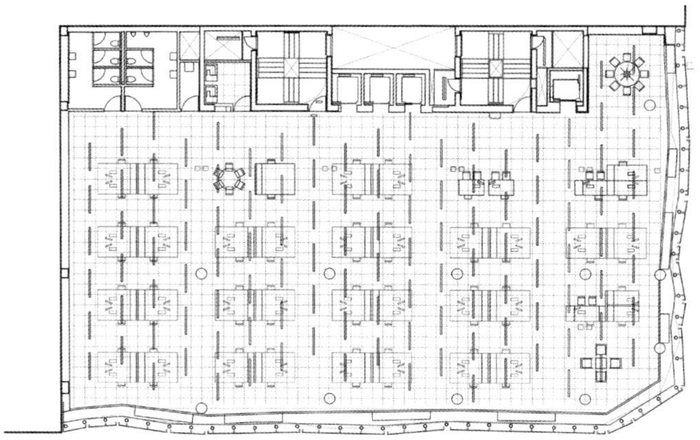

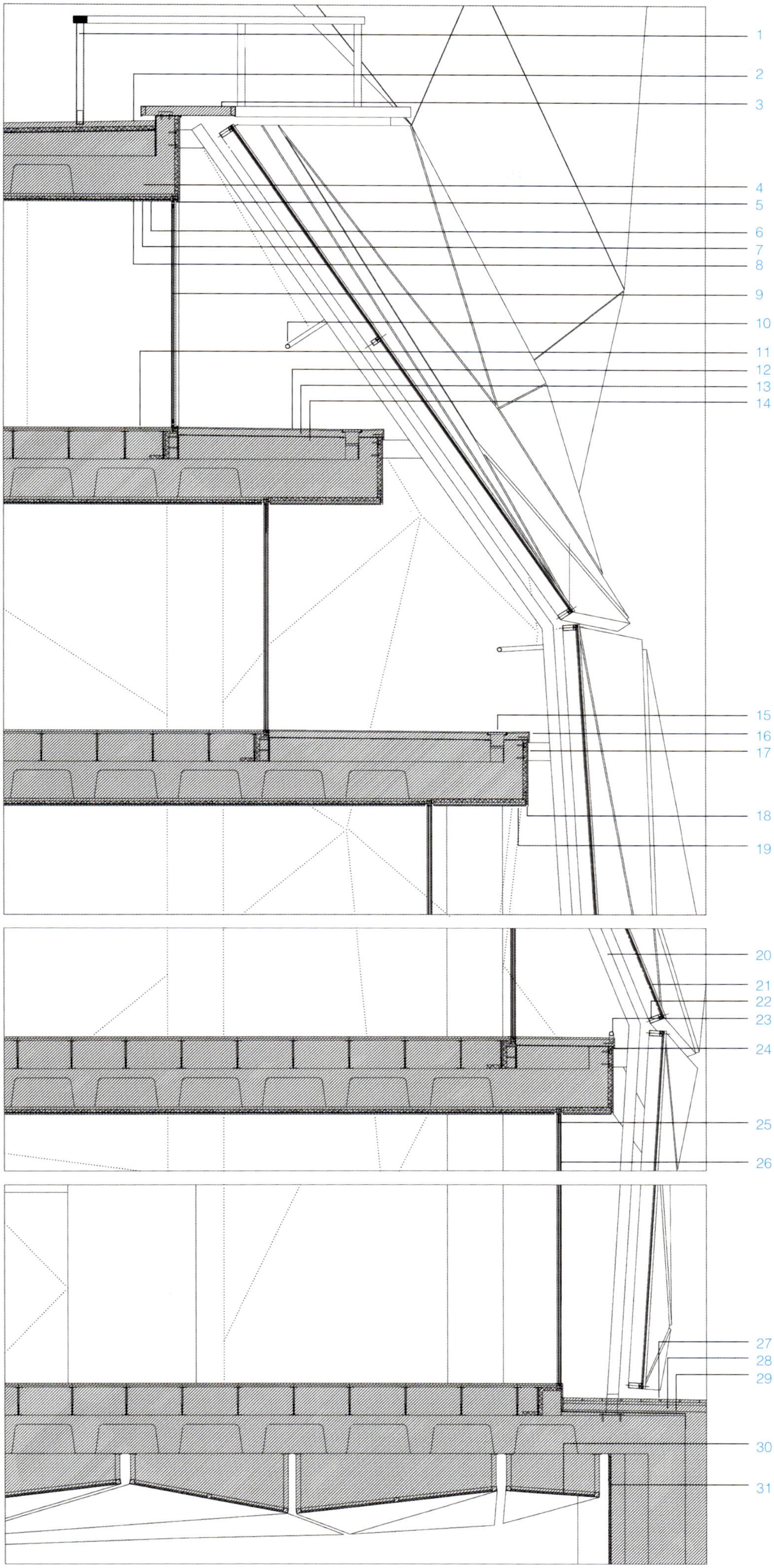

1. Banister: stainless steel circular tube profiles+wooden handrail
2. Reversed roof+reinforced concrete slab+finish polished
3. Ridge: aluminium folded metal sheets
4. Reinforced concrete waffle slab
5. Sliding doors: extruded anodised aluminium shapes
6. Thermal insulation
7. Radiant ceiling conditioning
8. Suspended ceiling: plasterboard sheets fixed to galvanised steel profiles
9. Low-E double glazing with interior security laminated glass
10. Handrail: stainless steel circular tube profile
11. Raised floor
12. Reinforced concrete slab
13. Waterproof membranes
14. Increased light mortar layer
15. Drain cover canal
16. 1/2 Ipe stainless steel profile
17. Front of structural floor: wood and cement
    Panel+mortar increased layer of special
    Waterproof cement+galvanised steel profiles+inside thermal insulation
18. Throat: folded aluminium sheets
19. Suspended ceiling: wood and cement panel mortar increased layer of special waterproof
    Cement+galvanised steel profiles+inside thermal insulation
20. Stainless steel tube profile stanchion
21. Laminated glass with sunlight control
22. Extruded anodised aluminium profile
23. Skirting board: stainless steel circular tube profile
24. 1/2 Ipe stainless steel profile
25. Extruded anodised aluminium carpentry with breaking of thermal bridge
26. Double low-E glazing with interior security laminated glass
27. Encaustic cement floor tile
28. Reinforced concrete slab
29. Waterproof membranes
30. Suspended ceiling: expanded stainless steel mesh+galvanised steel profiles+interior acoustic absorber
31. Interior coating: expanded stainless steel mesh+galvanised steel profiles+interior acustic absorber

1. 栏杆：不锈钢圆管型材+木制扶手
2. 翻转屋顶+抛光钢筋混凝土板
3. 屋脊：铝制折叠金属片
4. 钢筋混凝土井式楼板
5. 推拉门：挤压阳极电镀铝型材
6. 保温层
7. 辐射吊顶调节
8. 吊顶：石膏板固定在镀锌板塑钢型材之上
9. 低辐射双层玻璃，内置夹层安全玻璃
10. 扶手：不锈钢圆管剖面
11. 活动地板
12. 钢筋混凝土板
13. 防水卷材
14. 加强型轻质砂浆层
15. 排水管槽
16. 1/2工字钢不锈钢型材
17. 楼层结构的前端：木材和水泥
    板材+特殊加强型砂浆层
    防水水泥+镀锌钢板型材+内保温隔热层
18. 窄道：折叠铝板
19. 吊顶：木材、水泥板+特殊加强型防水砂浆层
    水泥+镀锌钢板型材+内保温隔热层
20. 不锈钢管侧面支柱
21. 能够有效控制光线照射的夹层玻璃
22. 挤压阳极氧化铝型材
23. 裙板：不锈钢圆管型材
24. 1/2工字钢不锈钢型材
25. 木料外覆以挤压阳极氧化铝涂层，与热桥断开
26. 双层低辐射玻璃，内置夹层安全玻璃
27. 上釉水泥地板瓷砖
28. 钢筋混凝土板
29. 防水卷材
30. 吊顶：延伸型不锈钢网+镀锌钢型材+室内吸音板
31. 室内涂料：延伸型不锈钢网+镀锌钢型材+室内吸音板

Bermuda Point Office Building

百慕大顶点办公大楼

Location: Queensland, Australia
Designer: Tony Owen Partners
Photographer: Brett Boardman
Completion date: 2006

项目地点：澳大利亚，昆士兰州
设计师：托尼·欧文伙伴设计公司
摄影师：布雷特·博德曼
竣工时间：2006年

Award name:
2007 Gold Coast Urban Design Awards

获奖名称：
2007年黄金海岸城市设计奖

Bermuda Point Office Building is a strata commercial building located opposite the Bond University on the Gold Coast. The building is located on the waters edge at Lake Orr and enjoys spectacular views across the lake to Arata Isozaki's university campus.

The developer's brief specified an economical building which would provide maximum flexible floor space for strata offices. So, whilst the building had to be an economical structure, it also had to satisfy a master plan requirement for the highest standard of design. The result was a building made of very robust materials, which takes a unique shape to create a very exciting design statement.

The site itself was problematic as the foundation material was very poor requiring deep piling under the entire structure. At the same time the site was subject to flooding so the building had to be raised one level above ground.

The design solution was to create a large simple floor plate that could be divided into smaller 'front to back' office suites. This involved dividing a 20metresx64metres floor plate into 8 strata tenancies of each 8metresx18metres in size. Each tenancy enjoys waterside frontage. A two-metre-wide open walkway at the front ensured each tenancy also had a street frontage.

The site itself is orientated north/south. In order to maximise views the long façades had to face east/west, and this created a sun control problem. At the same time, due to the flooding requirements, the building is supported above two levels of parking and it was a requirement of the master plan that the parking should not be visible from the university across the lake.

The building was cost-effective to build, using passive environmental principles combined with standard, robust materials. The portal, which unifies the building, is made from framed fibrous cement sheet, as are the louvre fins. The floor-slab soffits are painted concrete and the glazing is a standard aluminium style. The iconic shape combined with the use of the coloured fins allows this inexpensive building to make a strong, sophisticated design statement.

百慕大顶点办公大楼是一幢典型的多层商业建筑，位于黄金海岸邦德大学对面。该建筑坐落于奥尔湖畔，坐享湖边风光和由建筑大师矶崎新设计的校园景致。

项目的开发商要求建筑能够在节省运算的基础上提供一个灵活的办公空间。因此，设计师巧妙采用了一个经济结构，尽量满足总体规划中高标准设计的要求。建筑运用的材料较为天然、耐用，其构建出的独特建筑外观具有鲜明的个性，极具表现力。

因建筑所在地地形较为特殊，因此，整个建筑结构的下方打造了一个稳固的地基。同时，为应对洪水带来的威胁，建筑被抬高至地面一层楼的高度。

一个大型简约的楼板将办公空间巧妙地划分成小型“前后相通”的办公套间。20米x64米的楼板被分隔成8个租赁单元，每个单元面积为8米 x 18米。匠心独运的设计保证每个租赁单元能够尽享临湖美景。位于前方的一个2米宽的通道确保每个单元拥有正向临街的视角。

建筑呈南北朝向。为优化室内观景角度，狭长的立面采用东西朝向处理，能够对太阳光线进行有效调节。同时，考虑到洪水带来的困扰，建筑支持两层以上的停车位，而这也与总体规划中要求停车位不被湖对岸大学校园窥见的原则相得益彰。

建筑的设计符合成本效益，将被动环境原则与标准、耐用的材料相结合。完美衔接建筑的入口采用纤维水泥板框架结构，造型犹如百叶窗的散热片。楼板的底面覆以混凝土材料，规格的铝材令其熠熠生辉。别具一格的造型搭配“翅片”上绚丽的色调有力地传达出耀眼、精致之感。

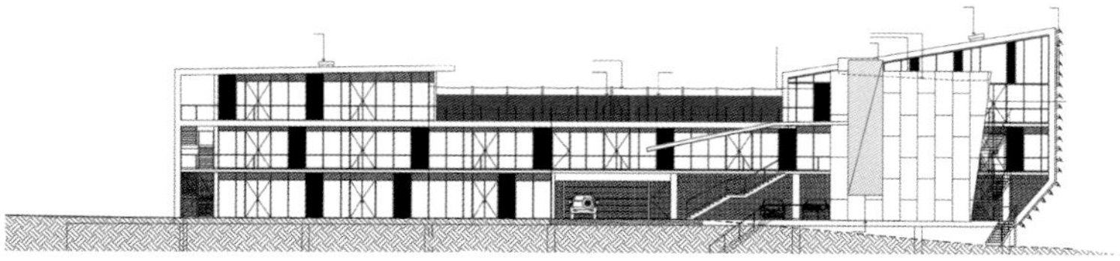

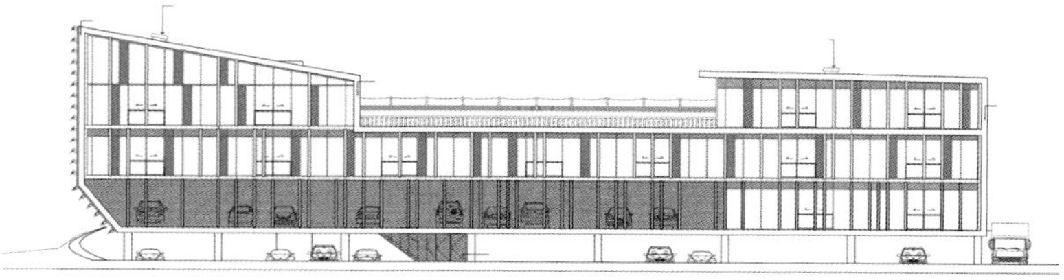

BERMUDA
POINT

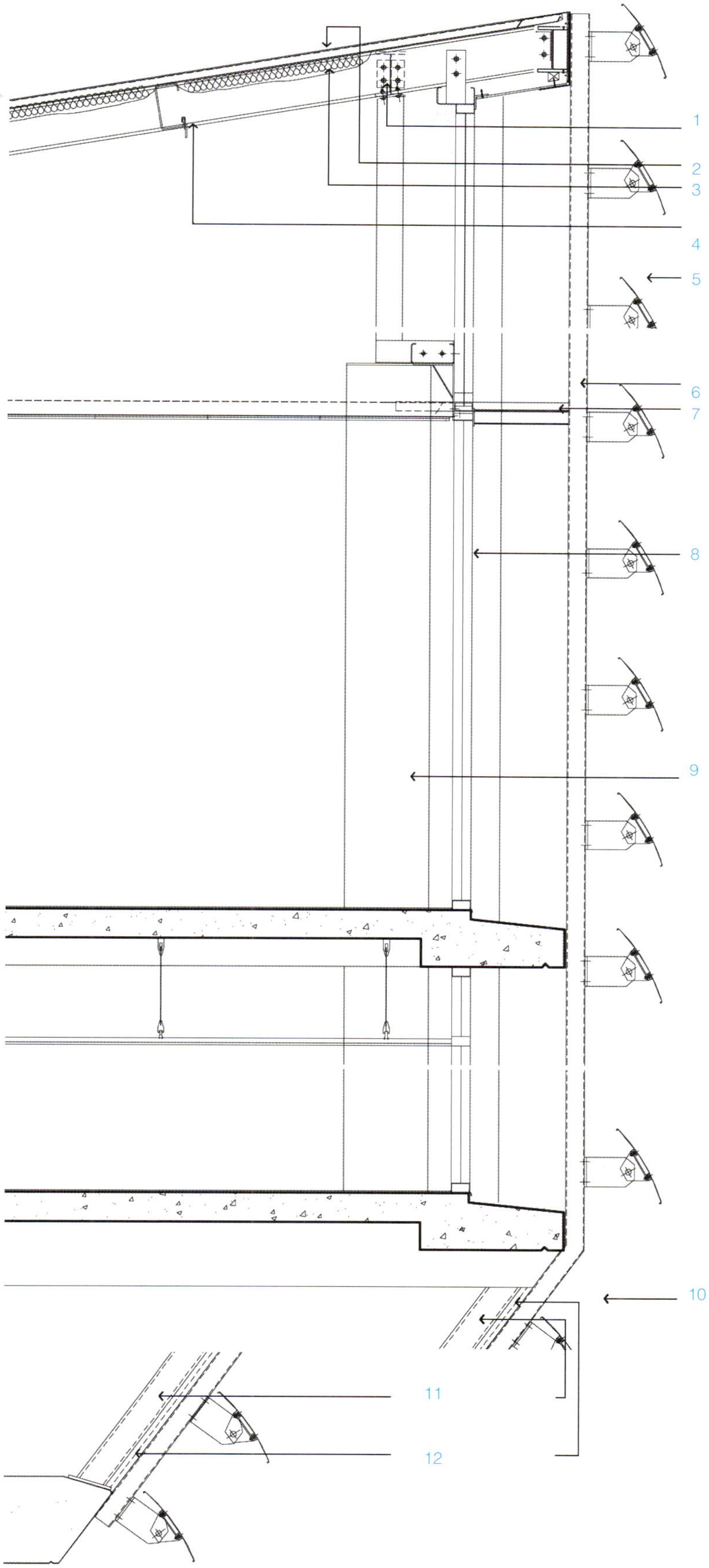

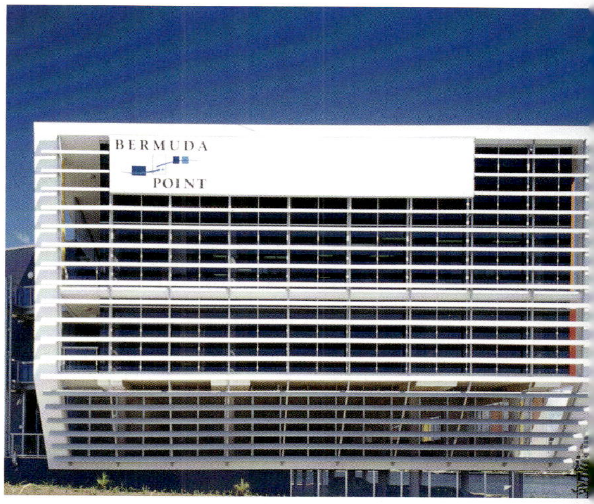

1. Steel roof framing
2. Colorbond klip-lok roofing
3. Insulation
4. Suspended ceiling+suspension system
5. Sunshade louvre blade device
6. 75 SHS to struct
7. Steel tie to struct
8. Glazing
9. HARDI column filled with concrete
10. Sunshade louvre blade device
11. Steel framing
12. Steel mesh

1. 屋顶钢架结构
2. 彩钢博思格来实KLIP-LOK金属屋面板
3. 绝缘层
4. 吊顶+悬垂系统
5. 遮阳百叶窗
6. 75毫米薄钢板结构
7. 钢材结构
8. 玻璃装配
9. 哈迪钢管混凝土柱
10. 遮阳百叶窗
11. 钢构架
12. 钢丝网

BERMUDA
POINT

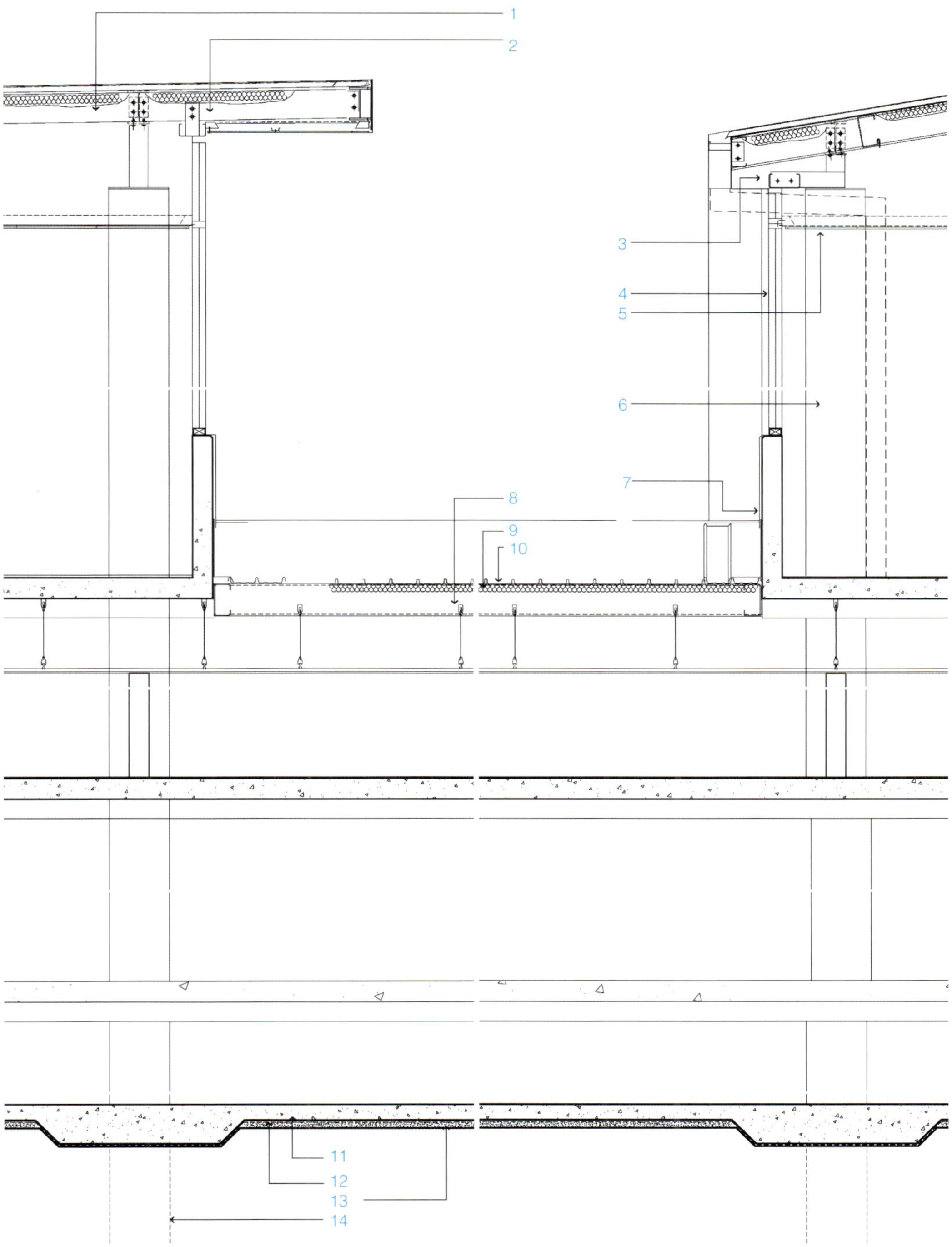

1. Steel roof framing
2. Insulation
3. Window head profile fixed to the top of column
4. Aluminium framed window
5. Suspended ceiling + suspension system
6. Hardi round columns filled with concrete
7. Flashing
8. Steel roof framing
9. Insulation
10. Colorbond klip-lok roofing
11. RC slab on ground
12. Vapour membrane
13. Sand bed
14. Pier as applicable

1. 屋顶钢架结构
2. 绝缘层
3. 窗头侧面被固定于廊柱的顶部
4. 铝合金窗框
5. 吊顶+悬垂系统
6. 哈迪填充混凝土圆形柱
7. 防水板
8. 屋顶钢架结构
9. 绝缘层
10. 博思格来实的KLIP–LOK金属彩钢屋面板
11. 钢筋混凝土板地面
12. 蒸汽膜
13. 沙床
14. 适用的窗间壁

1. Steel roof framing
2. Insulation
3. Colorbond klip-lok roofing
4. Obscure glass panel
5. Aluminium framed window
6. Suspended ceiling
7. 50millimetres setdown at RC slab edge
8. RC slab edge
9. Hardi round columns filled with concrete
10. Aluminium framed window
11. RC Slab thickening beyond
12. Suspended ceiling carpet
13. RC slab thickening beyond

1. 屋顶钢架结构
2. 绝缘层
3. 博思格来实的KLIP-LOK金属彩钢屋面板
4. 压花玻璃板
5. 铝合金窗框
6. 吊顶
7. 50毫米厚的钢筋混凝土板边缘
8. 钢筋混凝土板边缘
9. 哈迪填充混凝土圆形柱
10. 铝合金窗框
11. 超厚钢筋混凝土板
12. 吊顶覆盖物
13. 超厚钢筋混凝土板

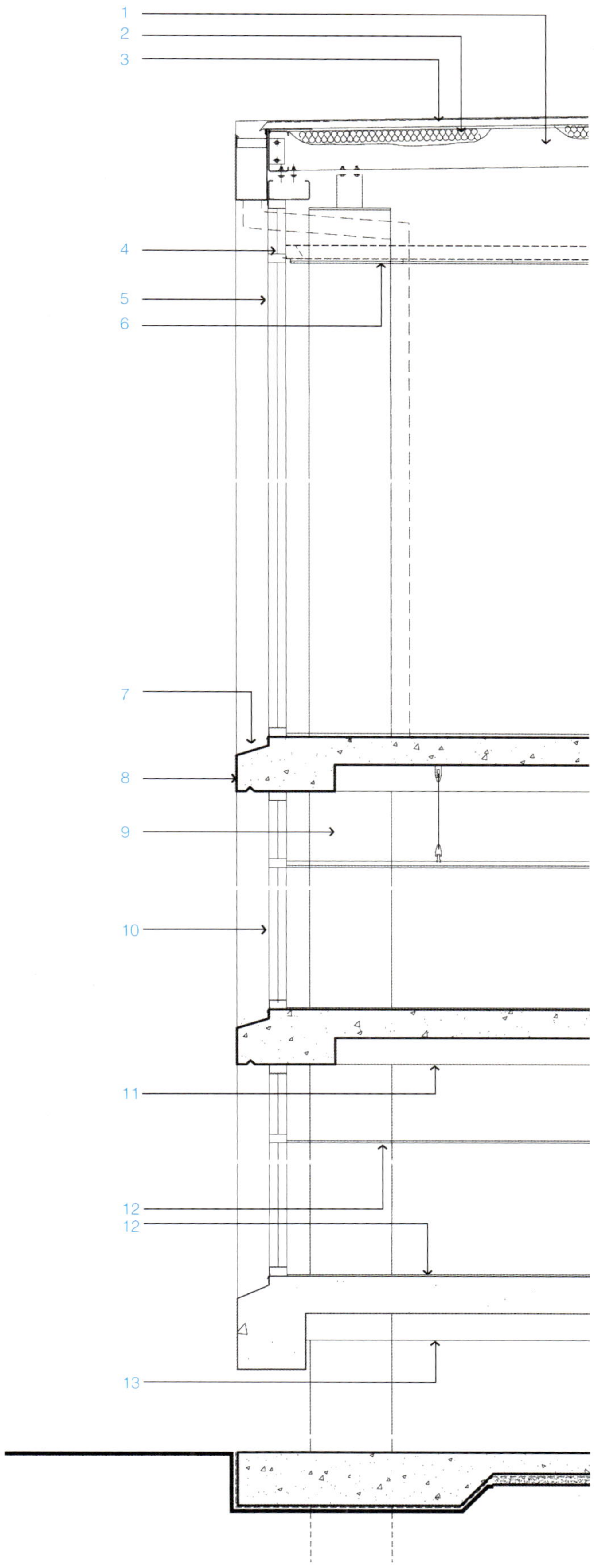

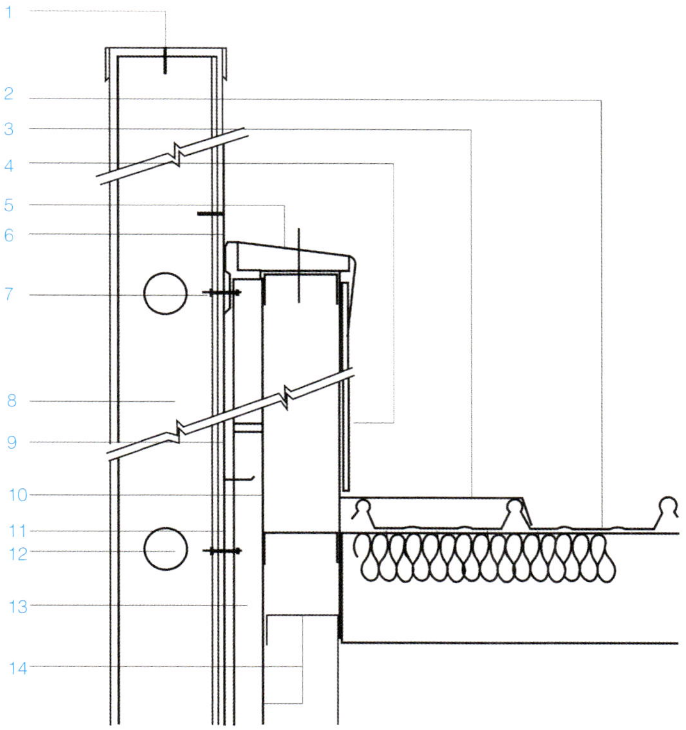

1. Capping piece
2. Klip-lok roofing
3. Flashing
4. FC sheet
5. Hardwood support
6. Screw fix capping and folded
7. Sheeting to backing strip over continuous bead of sealant
8. Signage panel
9. Exposed horizontal Joint 10mm NOM
10. Sarking
11. Express panel
12. Light
13. Top hat
14. Steel framing

1. 封盖件
2. 博思格来实的KLIP-LOK金属屋面板
3. 防水板
4. 光线通道板
5. 硬木支撑物
6. 螺丝封顶及褶曲
7. 密封剂上方的条状垫板挡板
8. 标牌板
9. 裸露的水平结合点10毫米（指定）
10. 屋面衬板
11. 挤压板材
12. 光线
13. 顶帽
14. 钢构架

Schwandorf

Schwandorf行政大楼

Location: Stuttgart, Germany
Designers: Archimedialab
Photographers: Happy Lim Photography
Completion date: 2007

项目地点：德国，斯图加特
设计师：Archimedialab建筑师事务所
摄影师：快乐林摄影工作室
竣工时间：2007年

Award name:
2007 Zuiderkerkprijs Award

获奖名称：
2007年Zuiderkerkprijs奖

The task to design a new administration building, reorganise the power station compound and create a new noise protection barrier offers the chance to dissolve the dichotomy of landscape and building to realise the deconstruction of those categories into one designed environment, to be experienced in a dynamic and curious fashion.
450 metres long and up to 13 metres high, the central part of a noise protection wall with a 45 degree incline simultaneously constitutes a new administration building for over 140 metres. The superimposition of building and earth wall allows one to explore and experience the landscape of this entire ensemble on various levels. An auditorium with a visitor centre unfolds from this landscape and opens up towards the power station compound. It separates from the earth wall on the upper level, resting on two radial supporting walls and cantilevers up to 20 metres over the landscape. A long panoramic glass façade leans towards the power station. The administration building underneath was sculpted into the earth wall. The distinctive shape of the auditorium is visible from far beyond the barrier, heralding a strangely new and positive identity for the place and the enterprise.
As for the part of the building that is situated within the core of the earth wall, a basic structure of exposed concrete was supplemented with built in furniture elements in coloured MDF, wooden oak floors, and brightly painted magnesia-bonded panels for acoustic absorbtion on walls and door elements. Part of the exposed concrete is stained in bright colours to contrast the archaic look of the untreated concrete. Profiled glass walls allow daylight to penetrate far into the building. All meeting and conference rooms were enclosed in frameless glass walls. The light timber structure of the upper glue-lam shell remains exposed, the diamond shaped curvilinear spaces between them were filled with acoustic panels stained in a dark purple colour.
A specially engineered and carefully balanced mixture of earths with fine aggregate and a very low percentage cement for adhesion ensures the stability of the earth wall structure, yet allows enough water to be retained to let vegetation cover the structure completely and evenly. A large portion of the building is covered with up to 6 metres of the earth wall. The upper floor of the building folds out of the wall and cantilevers up to 20 metres over the upper level of the site. A pre-stressed concrete structure, initially supported by two curvilinear walls allows the auditorium and visitor centre to hover over the site to allow panoramic views of the entire area.

该项目旨在设计一个全新的行政大楼，重新为发电厂内建筑进行布局，并通过一个隔音墙的构建，将建筑与景观综合一起，打造一个别具一格的动感空间。
隔音墙长450米，高13米，中央部分的倾斜角度约为45度，同时延伸出一个140多米长的新行政大楼。行政大楼和隔音墙的这种叠合设计，成功地将建筑和景观相结合，使人们在不同的楼层均能够将秀美的景致尽收眼底。附有访客中心的大礼堂与发电站相对，和绿色墙壁完全分开，并依赖两个径向支撑壁悬浮在空中，悬空部分大约长20米。通过大礼堂的全景玻璃立面，能够饱览发电站的全景。整个行政大楼都在隔音绿色墙壁之下。由于大礼堂的建筑外形非常独特，而且没有被绿色墙壁覆盖，因此格外醒目。
绿色隔音墙下的建筑内别有一番意境：裸露的混凝土基础结构、有色的中密度纤维板、橡木地板、墙壁和门上明亮的氧化镁镀层(能起到吸收声音的作用)等等处处彰显了设计师的细致用心。部分混凝土墙壁上的涂色醒目而耀眼，与其他未加处理的混凝土墙形成鲜明的视觉对比。玻璃幕墙能够将自然光线自然引入室内。无框架的玻璃幕墙将所有的会客室和会议室紧紧环绕。屋顶上的轻质木板结构也未经过任何处理，木板结构中间的菱形部分装有深色的隔音板。
经过谨慎周密的研究以及对泥土、细集料和少量黏合水泥的百分比的均衡考虑，设计师最终构建出这个牢固的墙壁结构，而该结构还可以吸附部分水分，促进植被更好地生长。除了建筑的高层部分，建筑的大部分均被这种绿色墙壁所覆盖。建筑的高层部分，也就是大礼堂所在的楼层，悬浮在空中，伸出去约20米。它是一个预应力混凝土结构，由两个径向承重壁支撑，能够将整个建筑的全景尽收眼底。

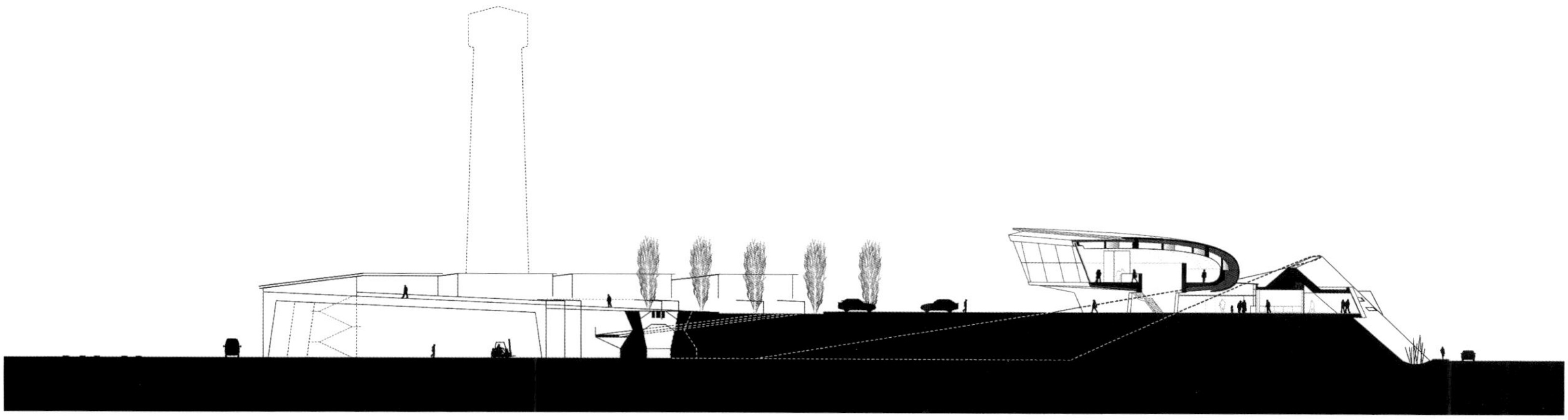

1. Lattice fresh air intake 1.35square metrex
   3 elements 0.38x1.20m
2. Multi-layered panels d = 60
3. Level sun and glare protection outside
4. Aluminium façade posts T-section
5. Interior façade element

1. 通风口格栅 1.35平方米
   3 元素0.38x1.20m
2. 多层板d=60
3. 外部遮阳遮光层
4. 幕墙立柱T型截面
5. 内墙元素

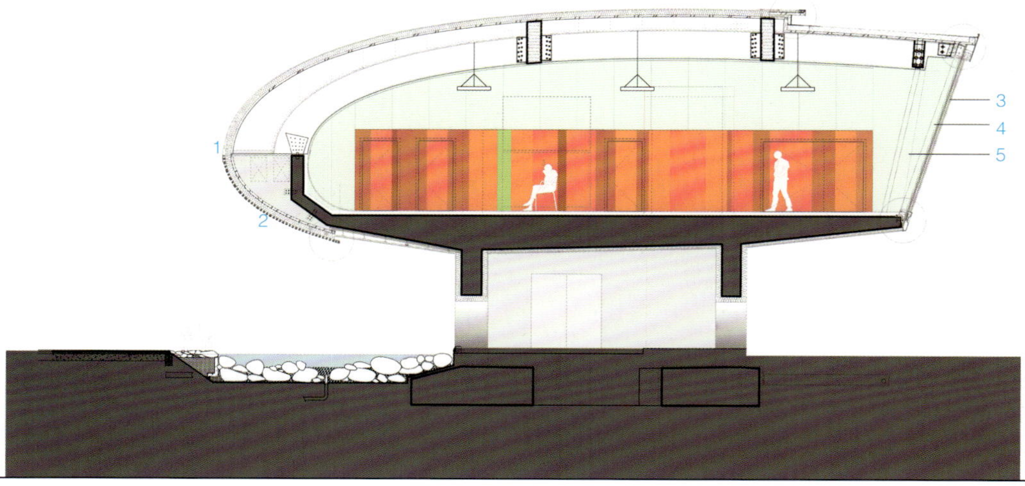

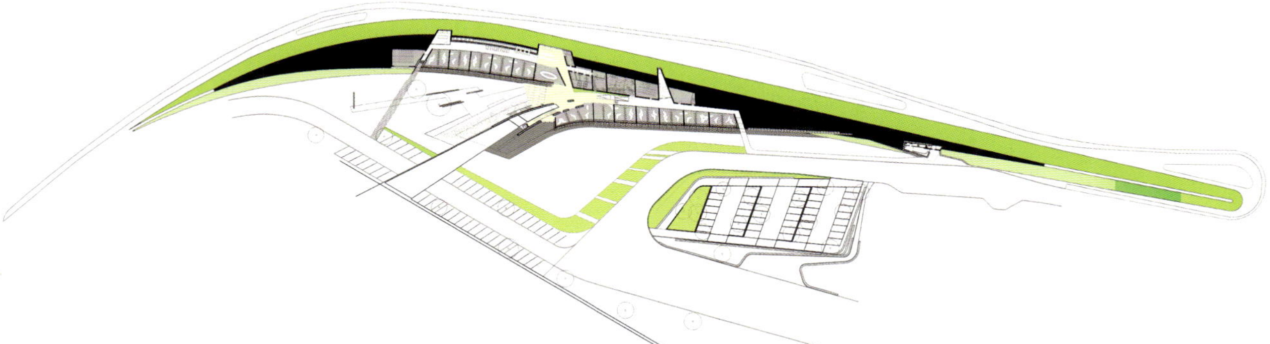

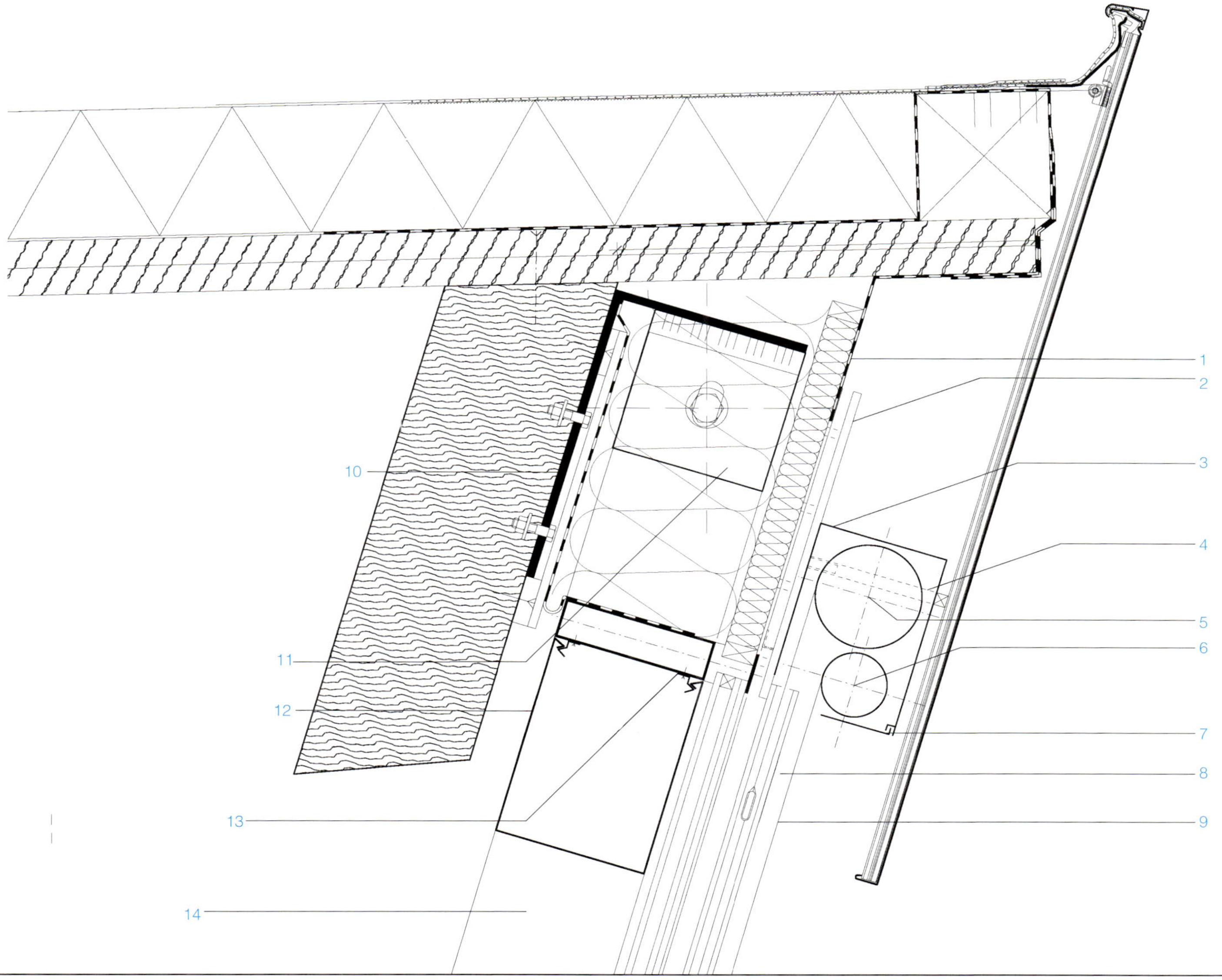

1. Panel
2. Bracket for the guide rails, the box of the sunshade and attica plate fastend to façade posts by thread bolt
3. Textile sun protection as countertension system Aluminium box about 160mmx230mm fastend to façade posts by thread bolt
4. Grouted anchor attica plate on thread bolt at bracket
5. Roller shaft
6. Motor shaft, cable inlet lays hidden in façade profile and to the junction box
7. Inspection cover removable from below
8. Guide rails for sunshade
9. Cord of the countertension system
10. Steel angle fastend with screws on glulam beam
11. Façade posts fastend by bolts between two steel plates with oblong holes for the vertical movements. The steel plates have 20mm distance to the posts, so that horizontal displacement also has space
12. Aluminium cover plate between the façade posts
13. Beam box section, altitude varies
14. Posts box section, inclination 16.65° to the vertical

1. 板材
2. 用于导轨、遮阳盒以及屋顶盖板的支架通过螺栓固定在幕墙立柱上
3. 织物遮阳作为反作用设备
大约160x230mm的铝盒通过螺栓固定在幕墙立柱上
4. 屋顶盖板锚杆灌浆在支架的螺栓上
5. 滚轮轴
6.电机轴，电缆入口隐藏在幕墙立柱中并与室内的接线盒相连
7. 检查盖可从底部拆卸
8. 遮阳导轨
9. 反作用系统缆绳
10. 角钢通过螺丝固定在胶合层积材梁上
11. 幕墙立柱通过螺栓固定在有长圆孔的2张铁板之间并可垂直移动。铁板与立柱之间各有20mm的距离，所以平行移动也可
12. 铝盖板在幕墙立柱之间
13. 箱型梁截面，高度不同
14. 箱型柱截面，向垂直方倾斜16.65°

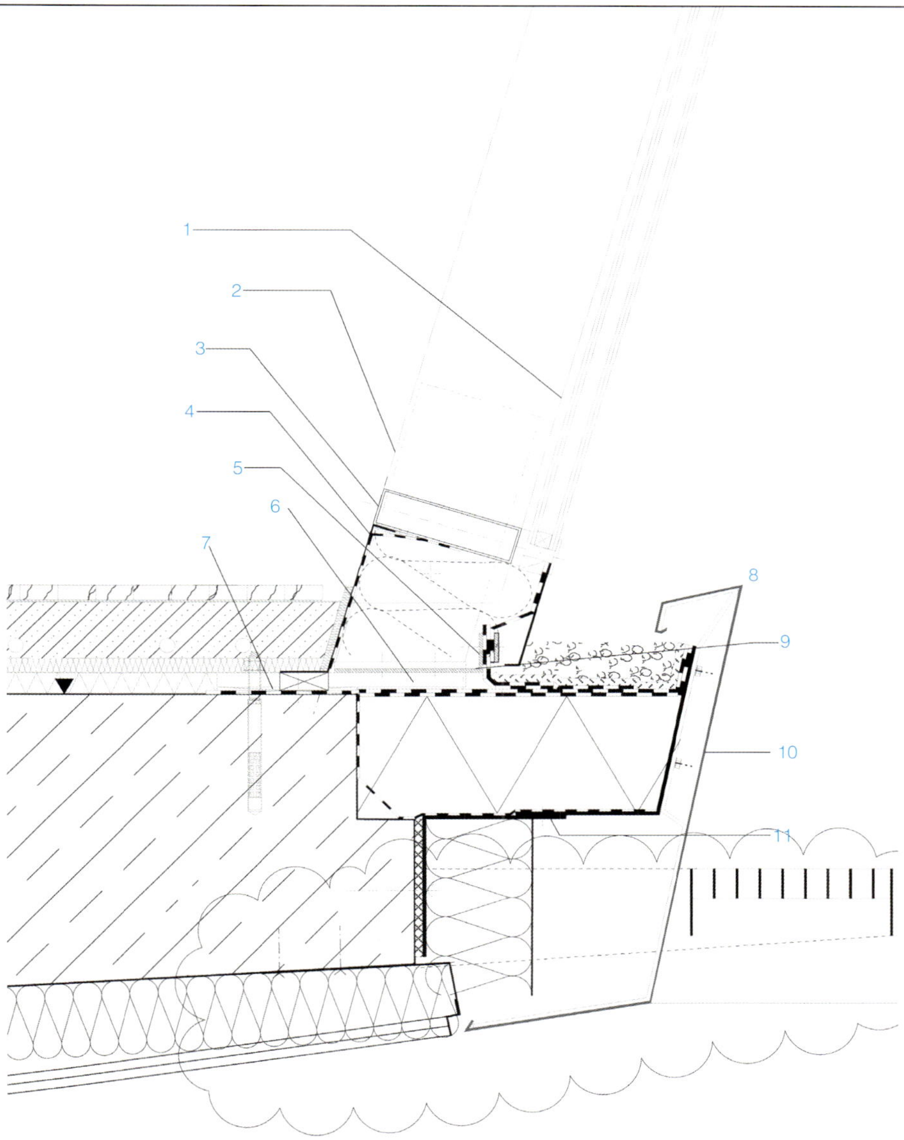

1. Over head and safty glazing from the inside outwards
2. Façade posts
3. Façade beam
4. Facing invisibly mounted
5. Frame with loose-/fast-flange connection
6. Steel plate
7. Height compensation
8. Roof structure:
   50mm gravel
   10mm builds protection mats
   bituminous roof-sealing, 2-ply
   160mm insulation
   bituminous vapour barrier
   Undercoating
9. Sealing connection after frame installation
10. Attic covering on bracket
11. Attic subconstruction

1. 吊顶安全玻璃由内到外
2. 幕墙立柱
3. 幕墙横梁
4. 饰面,不可见固定
5. 松紧法兰连接的边框
6. 钢板
7. 高度调整
8. 屋顶结构：
   50mm砾石
   10mm建筑防护垫
   2层沥青屋顶密封
   160mm绝缘层
   沥青水汽隔绝
   防水底涂层
9. 框架安装后密封接口
10. 支架上的顶楼包面
11. 顶楼底部结构

1. Reinforced concrete sloped ceiling
2. Undercoating
3. 2-layer waterproofing
4. 14cm thermal insulation
5. Trickle protection mat
6. 5cm gravel
7. Shear threshold structure schubschwellenkonstruktion
8. Reinforced concrete
9. Undercoating
10. 2-layer waterproofing
11. 14cm thermal insulation
12. Trickle protection mat
13. 5cm gravel
14. Ok RD im gefalle 2%

1. 钢筋混凝土坡顶
2. 防水底涂层
3. 二层防水
4. 14cm保温隔离
5. 涓流保护毡
6. 5cm砾石
7. Shear threshold structure schubschwellenkonstruktion
8. 钢筋混凝土
9. 防水底涂层
10. 二层防水
11. 14cm EPS保温隔离
12. 涓流保护毡
13. 5cm砾石
14. Ok RD im gefalle 2%

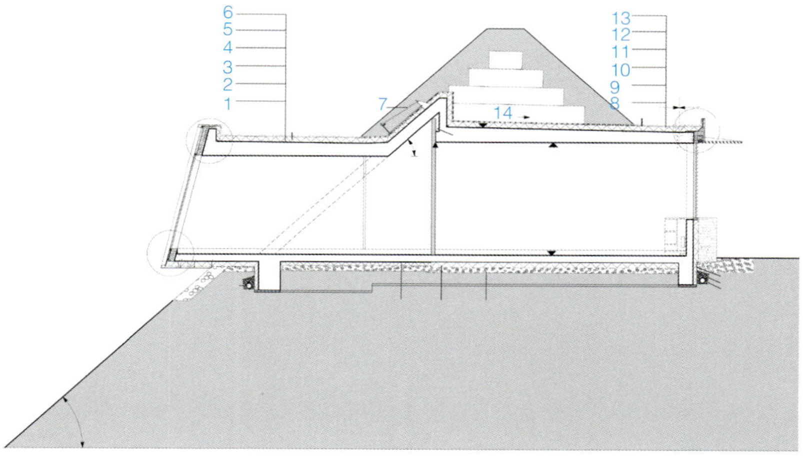

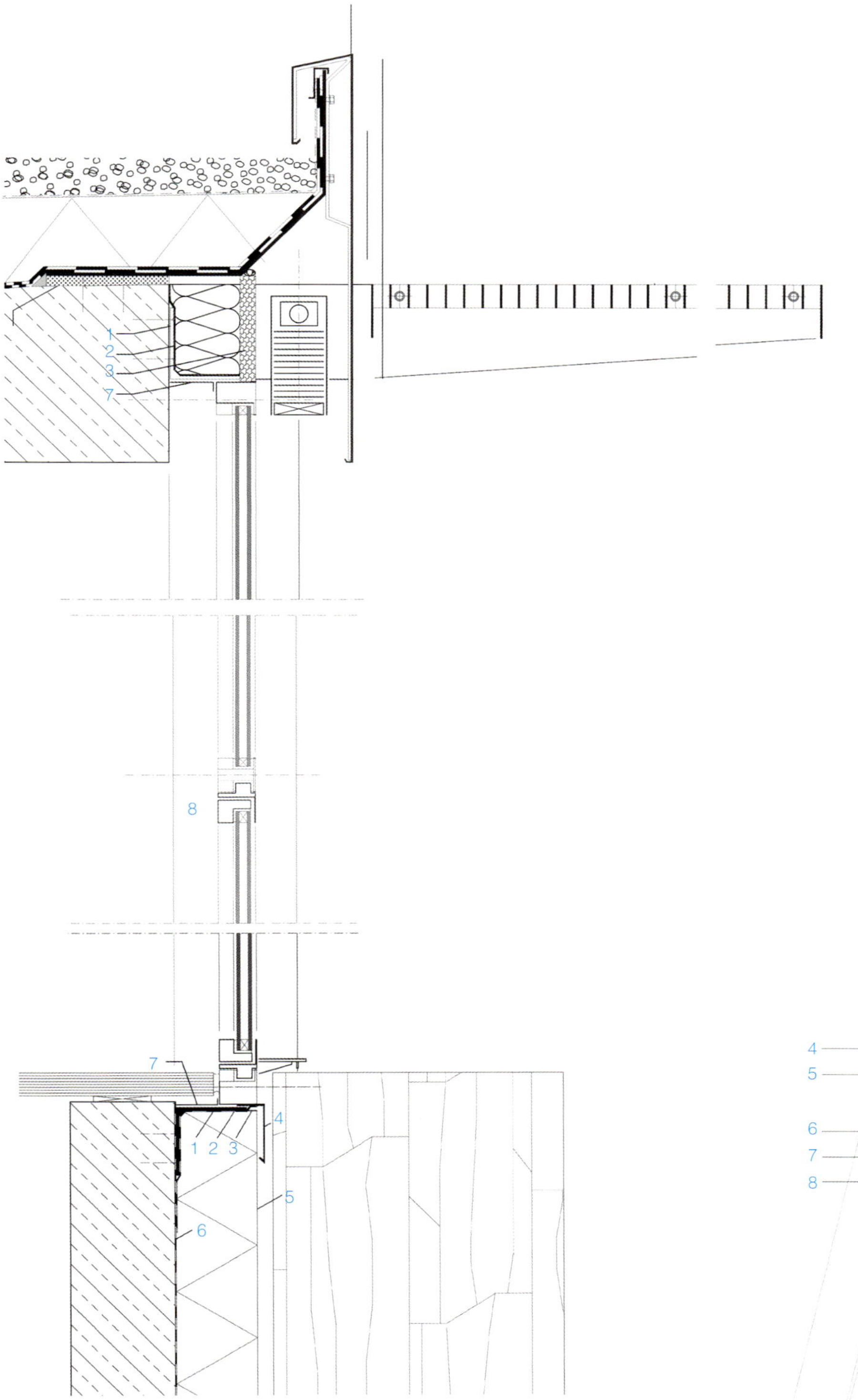

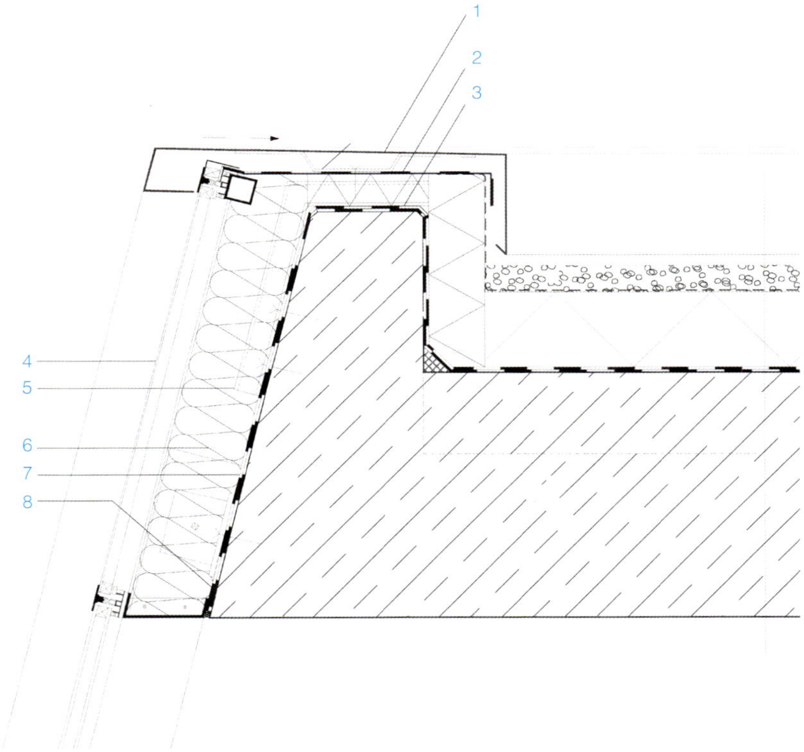

1. Subconstruction at concrete parapet or ceiling front edge, fastened façade and balancing of tolerance
2. Vapour proof connection at concrete Parapet
3. Cold bridge-free connection hollows stuffed with mineral fibre
4. Aluminium cover plate, thickness 2mm, powder coated surface, drainage behind cover plate
5. Perimeter insulation
6. Sealing of the concrete parapet
7. Continuous aluminium cover plate
8. Vertical connection between the façade profile and concrete wall

1. 混凝土护墙或顶板边缘的底部结构用于固定立面及平衡误差
2. 混凝土护墙上的水汽隔绝连接
3. 无冷桥连接，凹陷处用矿棉填充
4. 连续铝制盖板，厚度2mm，表面粉末涂层，盖板后的排水槽
5. 维护结构保温
6. 混凝土护墙密封
7. 连续铝制盖板
8. 幕墙立柱及混凝土墙垂直连接

1. Aluminium attica plate surface powder coated, fixed by adhesion at angled brackets
2. Sealing foil in façade profile clamped
3. Sealing
4. Opaque glazing panel
5. Angled brackets (Z-section) attached to the shell construction, centre distance about 500 mm
6. Thermal insulation mineral fibre
7. Vapour barrier on concrete
8. Vapour proof connection

1. 铝制屋顶边缘盖板，直径3 mm表面粉末涂层，通过角型托架固定
2. 密封薄膜夹在幕墙立柱中
3. 密封
4. 不透明玻璃嵌板
5. 角型托架（z形截面）固定在外壳上，中心距大约500 mm，螺钉穿透封闭
6. 保温矿棉
7. 混凝土的水汽隔绝
8. 水汽隔绝连接

Nestle Application Group

雀巢咖啡公司大楼

Location: Querétaro, Mexico
Designers: Rojkind Arquitectos
Photographers: Paúl Rivera
Completion date: 2009

项目地点：墨西哥，克雷塔罗
设计师：Rojkind建筑师事务所
摄影师：保罗·里韦拉
竣工时间：2009年

Award name:
The Best Renovations of 2009

获奖名称：
2009最佳革新项目奖

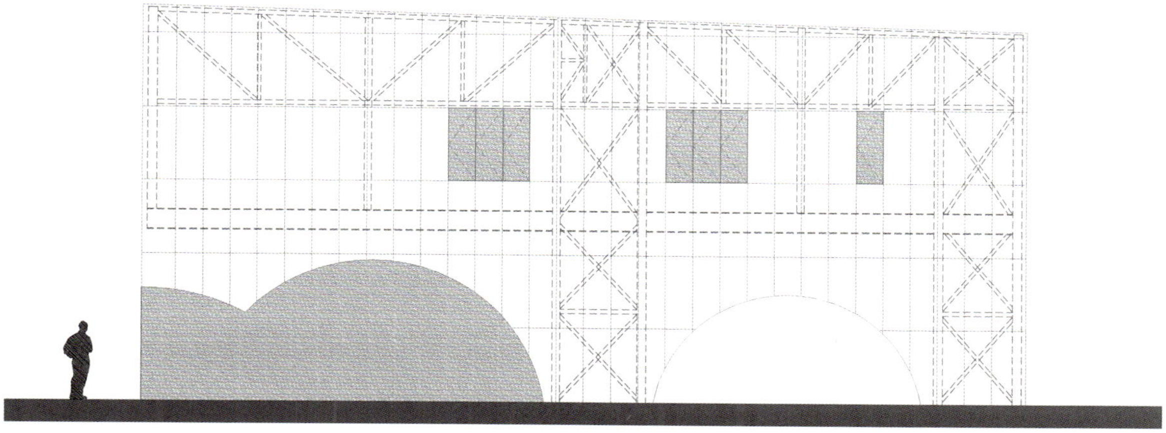

The UNESCO's designation of Queretaro's historic city centre as a world heritage site in 1996, had unforeseen consequences that even expanded to the city's industrial periphery. As a result of this designation, the new building was to have an arched porch, as rooted in tradition. Rojkind responded to this challenge with a reinterpretation not only of the arch but also of the porch. If the arch is nothing else than a fragment of a cupola, in the same vein, the cupola is an amplified arch when it rotates around its own axis, the cupola meets the reference criteria of the arch without turning it into a cliché.
While the exterior is opaque, metallic and impenetrable in appearance, the interior of these boxes painted in different colours, have an almost theatrical quality to them: it appears as if the researchers wearing their white robes were floating in a continuous flow of blues, yellows or greens which are interrupted by the continuous space of different colours sometimes, when one of the metal panels that covers the boxes reveals itself and opens like a window, which can be seen from the outside. The construction of these buildings (if built in a different latitude, a more sophisticated technology most likely would have had been employed to automatise the production of the unique geometries of these spheres) implied the translation of spatial forms was to be done in a different constructive manner, in a simple almost colloquial way, which allowed the local workers to fabricate the foam like spherical space from the physical intersection of the spherical cupolas made of rebar rings and arches.

1996年，联合国教育科学文化组织将墨西哥盖莱达洛市中心列为世界文化遗产，要求新建筑要拥有一座拱形门廊，这也给建筑师的设计带来一些约束。Rojkind建筑师事务所的新设计对拱门和走廊进行了完美诠释。他们通过挖掘建筑的直角创建一系列球体交叉领域，这些球体巧妙打造了一个开放、连续的拱形空间。这种设计既符合了建筑要求，而又不落俗套，匠心独运。
不透明的建筑的外观采用金属等材料，防风、隔热，为室内环境提供有效的保护。空间内部颜色富于变化， 五彩斑斓。穿着白袍的研究人员穿梭于空间内，仿佛漂浮于蓝色、黄色、绿色的连续气流之中，带给人全新的视觉享受。空间外围的金属板窗口式设计，将自然光线成功引入到室内中，时刻确保室内通透、明亮。
建筑的结构（或许在其他的项目中可能要采取更为先进的技术打造这些造型独特的几何形状球体）采用了简约、传统的设计方式，仅仅依靠当地工人利用钢拱支架及吊环而完成。

1. Convention room
2. Access
3. Packaging laboratory
4. Project leaders
5. Laboratory
1. 会议室
2. 入口
3. 包装实验室
4. 项目负责人
5. 实验室

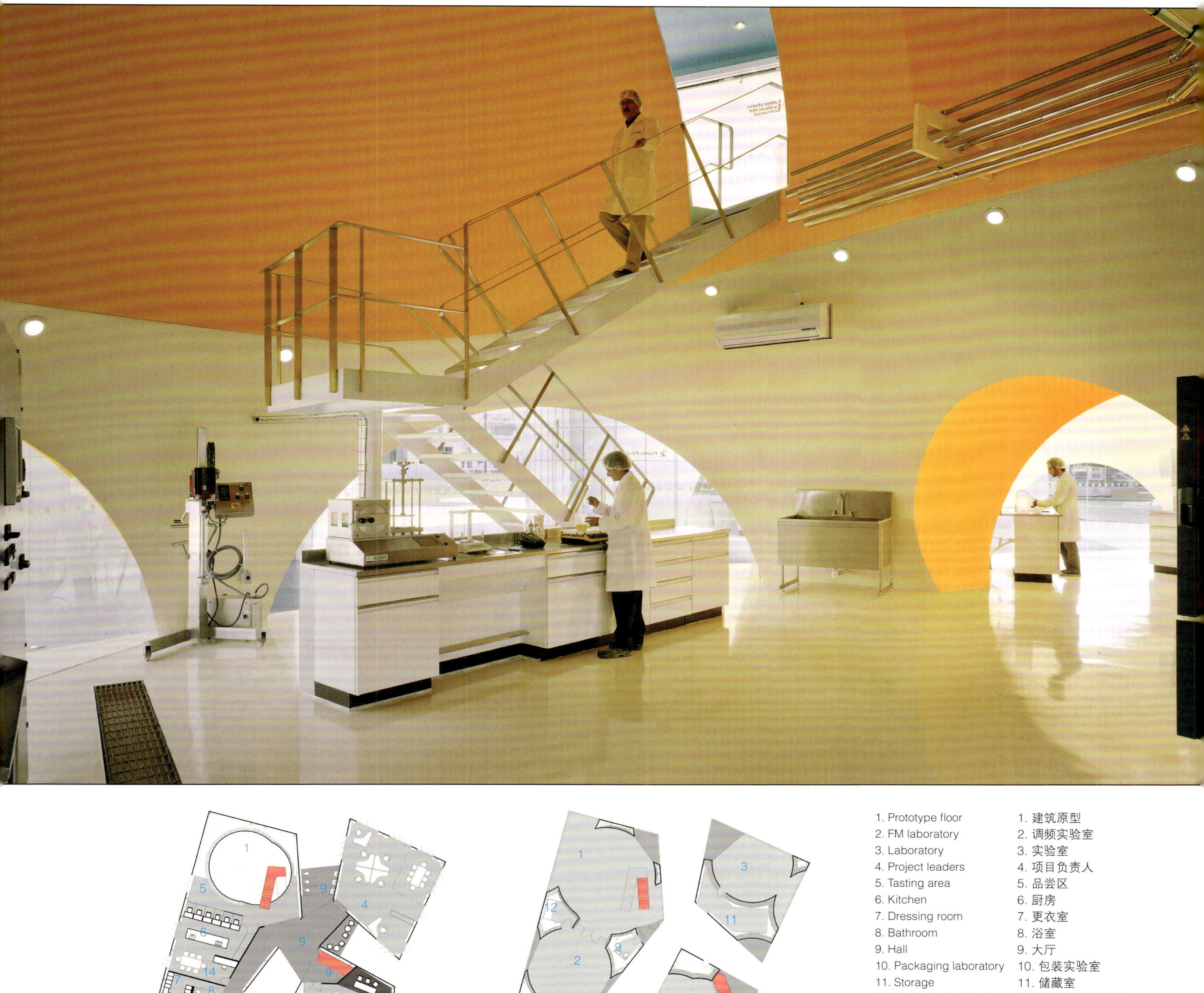

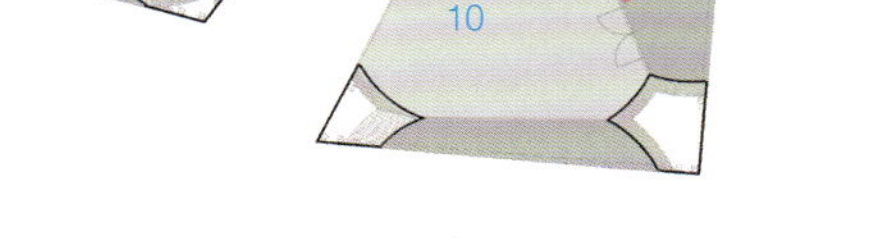

1. Prototype floor	1. 建筑原型
2. FM laboratory	2. 调频实验室
3. Laboratory	3. 实验室
4. Project leaders	4. 项目负责人
5. Tasting area	5. 品尝区
6. Kitchen	6. 厨房
7. Dressing room	7. 更衣室
8. Bathroom	8. 浴室
9. Hall	9. 大厅
10. Packaging laboratory	10. 包装实验室
11. Storage	11. 储藏室
12. Controlled ambient	12. 环境控制设备间
13. Access	13. 入口
14. Board room	14. 董事室

Nestle

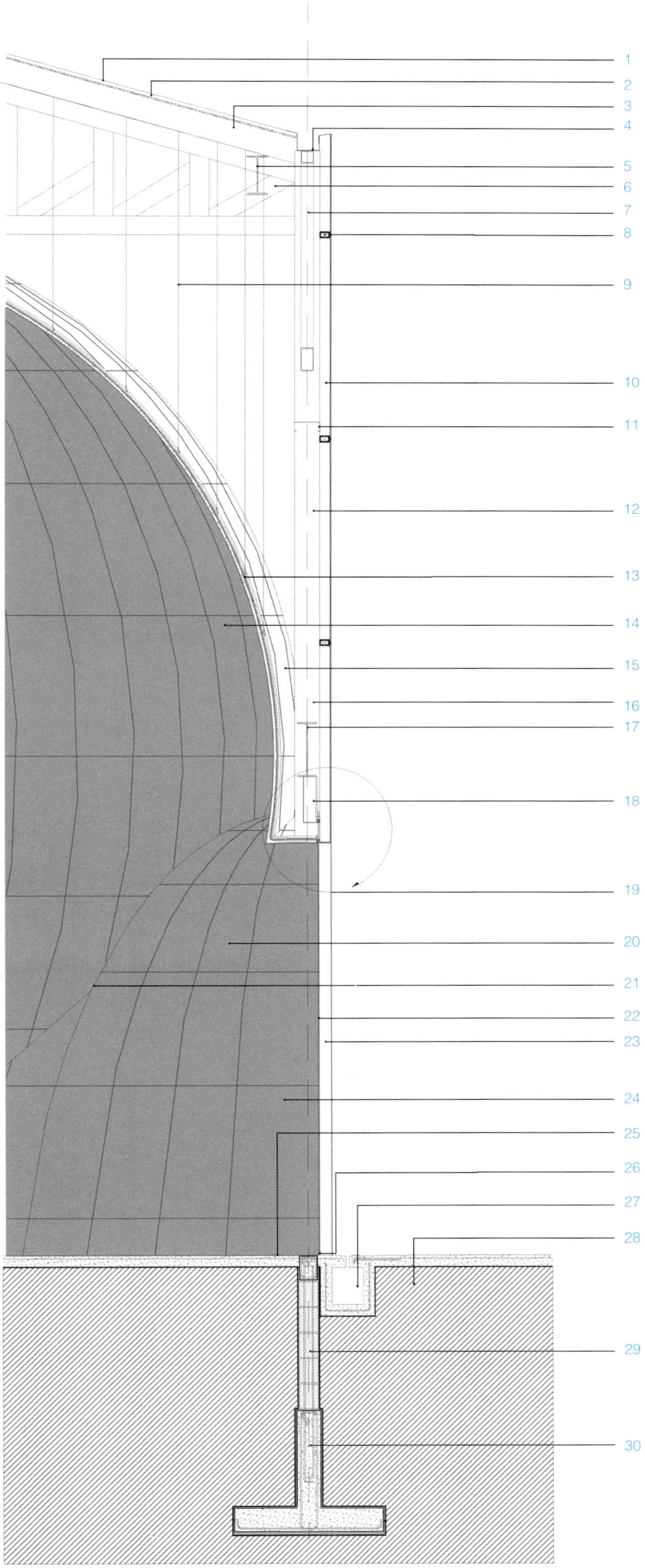

1. 13.5 mm cement board cover
2. "Thermafibre" (mineral wool) acoustic and thermal insulation
3. Mon-Ten 8inches structural post 7.88kilograms
4. Galvanised sheet gutter
5. 2x35x52.2kilograms WI steel beam
6. HSS profile frame
7. HSS profile frame
8. Tubular profile
9. Diameter wire spaced each 50 cm
10. Tubular profile
11. MON-TEN 8inches steel profile 10.2 kilogram/metre
12. 8inchesx8inchesx3.8 HSS profile 56.1 kilograms
13. Unfolded metal and sand cement flatter covered sphere structure
14. Epoxy paint finish according to approved sample
15. Sphere made of 3/8inches reinforcement steel grid
16. 8inchesx8inchesx3.8 HSS profile 56.1 kilograms
17. 8inchesx46cmx68.4kilograms WI steel beam
18. Supporting system for crystal made on site, perforated to receive ironwork
19. See Glass fixing detail in bulletin
20. Epoxy paint finish according to approved sample
21. Metallic plate used to cover intersection
22. 9mm tempered glass fixed with norton tape
23. Steel solepiece following façade's arch
24. Sphere made of 3/8inches reinforcement steel grid
25. Polished concrete firm f'c=200 kg/cm^2 reinforced with 6x6-10/10 electric welded mesh finished with epoxy paint
26. Stainless steel framework to receive glass
27. Concrete watercourse reinforced with 3/8inches spaced each 25 cm with integral waterproofing
28. 95% proctor compact ground
29. Foundation brick wall
30. Foundation wall footing

1. 13.5毫米水泥板盖
2. 矿物棉隔音和保温层
3. 20.3厘米Mon-Ten结构柱7.88千克
4. 镀锌板排水沟
5. 2x35x52.2千克熟铁钢梁
6. 高速钢侧面框架
7. 高速钢侧面框架
8. 管状剖面图
9. 线径间隔50厘米
10. 管状剖面
11. 20.3厘米Mon-Ten塑钢型材10.2千克/米
12. 20.3X20.3厘米x3.8高速钢侧面框架56.1千克
13. 折叠金属和砂石水泥平板覆盖球体结构
14. 根据核准的样本提供的环氧面漆
15. 球面由7.62 / 20.3厘米强化钢网制成
16. 20.3X20.3厘米x3.8高速钢侧面框架56.1千克
17. 20.3x46厘米x68.4千克熟铁钢梁
18. 现制晶体支承系统，穿孔铁制品加工
19. 第26号公告牌上的玻璃固定细节
20. 根据核准的样本提供的环氧面漆
21. 用于覆盖交叉点的金属板
22. 9毫米钢化玻璃由Norton胶布固定
23. 沿立面拱门的钢底板
24. 由7.62 / 20.3厘米强化钢网制成的球体
25. 抛光混凝土f'c=200千克/平方厘米，利用表面涂以环氧漆材料的电焊网6x6-10/10进行加固
26. 不锈钢玻璃构架
27. 混凝土水道由7.62 / 20.3厘米空间进行加固，每间隔25厘米设有整体防水系统
28. 95%普罗克特压缩地板
29. 地基砖墙
30. 基础墙基脚

Planta Piloto

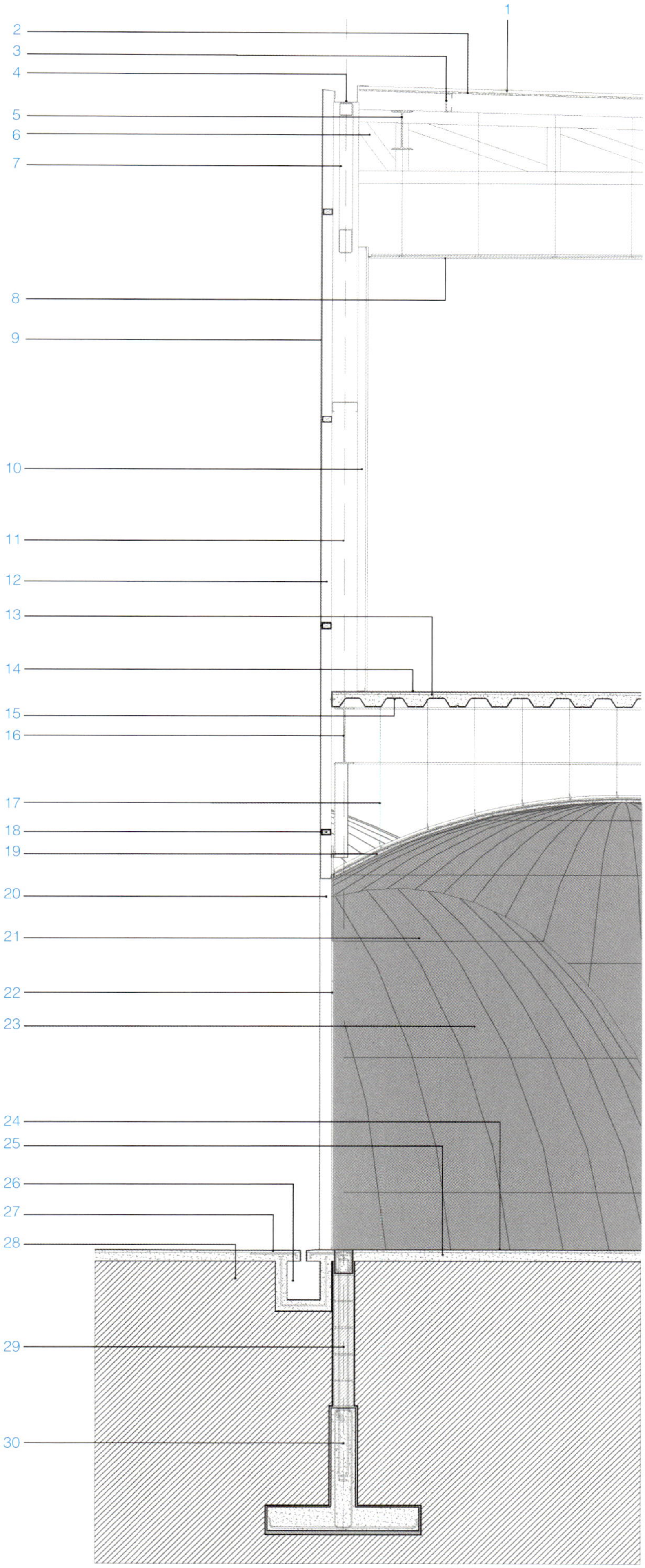

1. 13.5 mm cement board cover
2. "Thermafibre" (mineral wool) acoustic and thermal insulation
3. 6inches structural wood post 6.17 kilograms
4. Galvanised sheet gutter
5. 2x35x52.2 kg WI steel beam
6. HSS profile frame
7. HSS profile frame
8. 15.99 mm plaster board ceiling ready to receive paint
9. "mirror colour flex" 6mm glass
10. One faced wall made of 15.9mm thick plaster board finished to receive paint according to approved sample
11. 8inchesx8inches3.8 HSS profile 56.1 kilograms
12. Tubular profile
13. Polished concrete firm f'c=200kg/cm^2 reinforced with 6x6-10/10 electric welded mesh finished with epoxy paint
14. Interface flor&superflor carpet colour Indian spice
15. Galvadeck25 caliber 22 steel slab
16. 8'x46cmx68.4kilograms WI steel beam
17. 1/8" diameter wire spaced each 50cm
18. Tubular profile
19. Unfolded metal and sand cement flatter covered sphere structure
20. Aluminium solepiece following façade's arch
21. Sphere made of 3/8inches reinforcement steel grid
22. 9mm tempered glass fixed with norton tape
23. Epoxy paint finish according to approved sample
24. Epoxy resin floor colour according to approved sample
25. Polished concrete firm f'c=200 kg/cm reinforced with 6x6-10/10 electric welded mesh finished with epoxy paint
26. Concrete watercourse reinforced with 3/8inches rod spaced each 25cm with integral waterproofing
27. Concrete firm f'c=200 kg/cm, reinforced with 6x6-10/10 electric welded mesh gravel aggregate wash finished
28. 95% proctor compact ground
29. Foundation brick wall
30. Foundation wall footing

1. 13.5毫米水泥板盖
2. 矿物棉隔音和保温层
3. 15.2厘米结构木杆6.17千克
4. 镀锌板排水沟
5. 2x35x52.2千克熟铁钢梁
6. 高速钢侧面构架
7. 高速钢侧面构架
8. 预备着色的15.99毫米石膏板吊顶
9. 6毫米玻璃"彩色屈曲镜面"
10. 根据核准的样本喷涂的一个饰面墙，该墙面由15.9毫米石膏板制成
11. 20.3X20.3厘米3.8高速钢侧面56.1千克
12. 管状剖面
13. 抛光混凝土f'c=200千克/平方厘米，6x6-10/10电焊网，网面涂以环氧漆
14. flor&superflor地毯接触面，色彩具有印第安风情
15. Galwadeck25材料，管径22钢坯
16. 2.4米x46厘米x68.4千克熟铁钢梁
17. 直径 2.5/ 20.3厘米，线间距50厘米
18. 管状剖面
19. 折叠金属和砂石水泥平顶加盖球体结构
20. 沿立面拱门的铝底板
21. 球面由7.6 / 20.3厘米强化钢网制成
22. 9毫米钢化玻璃由Norton胶布固定
23. 根据核准的样本而采用的环氧面漆
24. 根据核准的样本为环氧树脂地板着色
25. 抛光混凝土f'c=200千克/厘米，6x6-10/10电焊网，网面涂以环氧漆
26. 钢筋混凝土水道7.6 / 20.3厘米杯杆，总体防水系统间距为25厘米
27. 混凝土f'c=200千克/厘米，表面采用砂石骨料进行抛光的电焊网进行加固
28. 95%普罗克特压缩地板
29. 地基砖墙
30. 基础墙墙基脚

Cocoon

蚕茧办公大楼

Location: Zürich, Switzerland
Designer: Camenzind Evolution
Photographer: Camenzind Evolution
Completion date: 2007

项目地点：瑞士，苏黎世
设计师：Camenzind Evolution设计师事务所
摄影师：Camenzind Evolution设计师事务所
竣工时间：2007年

Cocoon is located in Zurich's Seedeld district on a beautiful hillside, which enjoys excellent lake and mountain views. The location's distinctive flair stems from the exceptional park-like settings, a green oasis into which Cocoon snugly nestles. Flanked on three sides by mighty, age-old trees the elliptical structure reads as a freestanding sculptural volume that gracefully spirals up from the park. The stainless steel mesh enveloping the building combines visual privacy with restrained elegance, while establishing a strong and unmistakable presence.
The bold stand-alone building embodies an innovative conception of interior spatial organisation and interaction with the surrounding environment. In doing so, it caters for a wide variety of workplace and occupancy concepts. With its spiral massing, Cocoon may be conceived as a sort of "communication landscape" that creates a unique spatial configuration and working environment in a matchless setting. The stepped, upward-winding sequence of segments also shapes the character of the building interior. All spaces are arranged along a gently rising ramp, which wraps around a central, light-flooded atrium. The space planning concept dispensed with the traditional division into horizontal storeys in favour of a seemingly endless sequence of elliptical floor segments. By eliminating the usual barriers to communication, this generates a unique spatial experience and working environment that unlocks a host of intriguing possibilities for interaction and cooperation. The floorspace design is occupancy-neutral and provides for fully flexible partitioning together with the adaptability necessary to meet the shifting needs of future users. Together, the various elements-lifts, spiral ramp, segments and stairwell-constitute a clearly structured, versatile circulation system that provides for both the desired interaction and the necessary flexibility to accommodate alternative uses.
Internally, as the ellipses expand with each turn of the spiral, the skylight void opens up in a stunning spectacle. Externally, the building adopts the guise of a dynamic, upward-reaching sculpture. The dramatic atrium, with its wealth of internal visual links, generates a natural ambience conducive to communication and a sense of community.
The façade assembly consciously adds a note of subtlety and sophistication to the overall composition. The building is wrapped in a fine, almost scaly veil of stainless steel wire mesh. This curtain curls elegantly upwards in soft lines along the expanding spiral, its junction with the roof terrace accentuated by an open façade frame.

蚕茧总部办事处位于瑞士苏黎世塞费尔德地区的一座美丽的山坡上,拥有极好的山湖自然景观。周围绿意盎然的自然环境为其提供了得天独厚的地理优势。建筑三面被参天的古树进行环抱，椭圆形的结构宛如一个从地面上升起的盘旋而上的优雅雕塑。不锈钢网将建筑的外部进行包覆，在营造安全的室内环境的同时，流露出一种拘谨的优雅，同时也打造了一个硬朗、简洁的外观。
这个外形醒目而独特的建筑体现了一个创新的设计理念，使内部空间与周围环境实现良好的互动。这种做法，旨在满足工作场所和其他类似功能空间的多样性。根据这个螺旋的概念，蚕茧建筑可被看作是一种“通信景观”，创建一个独一无二、无可比拟的空间结构和工作环境。螺旋上升的楼梯踏步也塑造了建筑物内部空间的特色。所有的空间围绕着一个灯光辉煌的中庭，沿一个斜坡冉冉上升。这种连续的螺旋空间营运了一个独特的空间体验和工作环境，消除传统水平空间的分隔带来的沟通障碍，充分发挥了互动和合作的可能性。该建筑楼面设计具有良好的适应性，可以提供很灵活的分区，以满足未来用户的不断变化的需要。总之，各种不同的要素——电梯、螺旋坡道、隔断和楼梯井构成了结构清晰、多功能的流线体系，为交流、互动提供了可能，也可以为其他用途提供必要的灵活性。
在空间的内部，由于椭圆的延伸和螺旋体彼此之间的交错，使得从天窗倾洒下来的光在空间形成了一个异常美妙、令人叹为观止的景象。在外部，建筑犹如一个向上无限延展的雕塑，极具动态之美。戏剧性的中庭，用其内部丰富的视觉造型元素，营造了一种有利于沟通和社区意识的自然氛围。
正立面的装饰元素有意识地对微妙和复杂的整体组成进行了诠释。精致的鳞片状不锈钢丝网将建筑完整地包裹起来。这种金属外表皮，沿扩展的螺旋体优雅的向上延伸，与屋顶露台相连接，衔接处因开放的立面结构而十分引人瞩目。

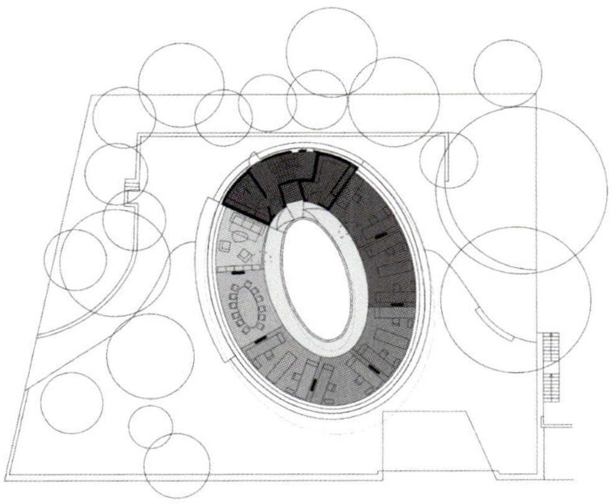

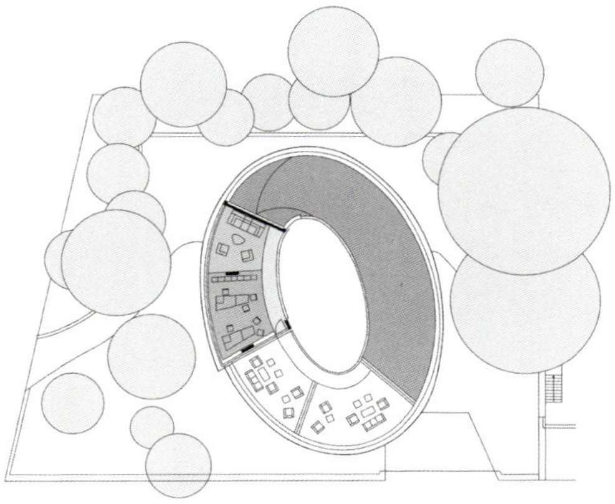

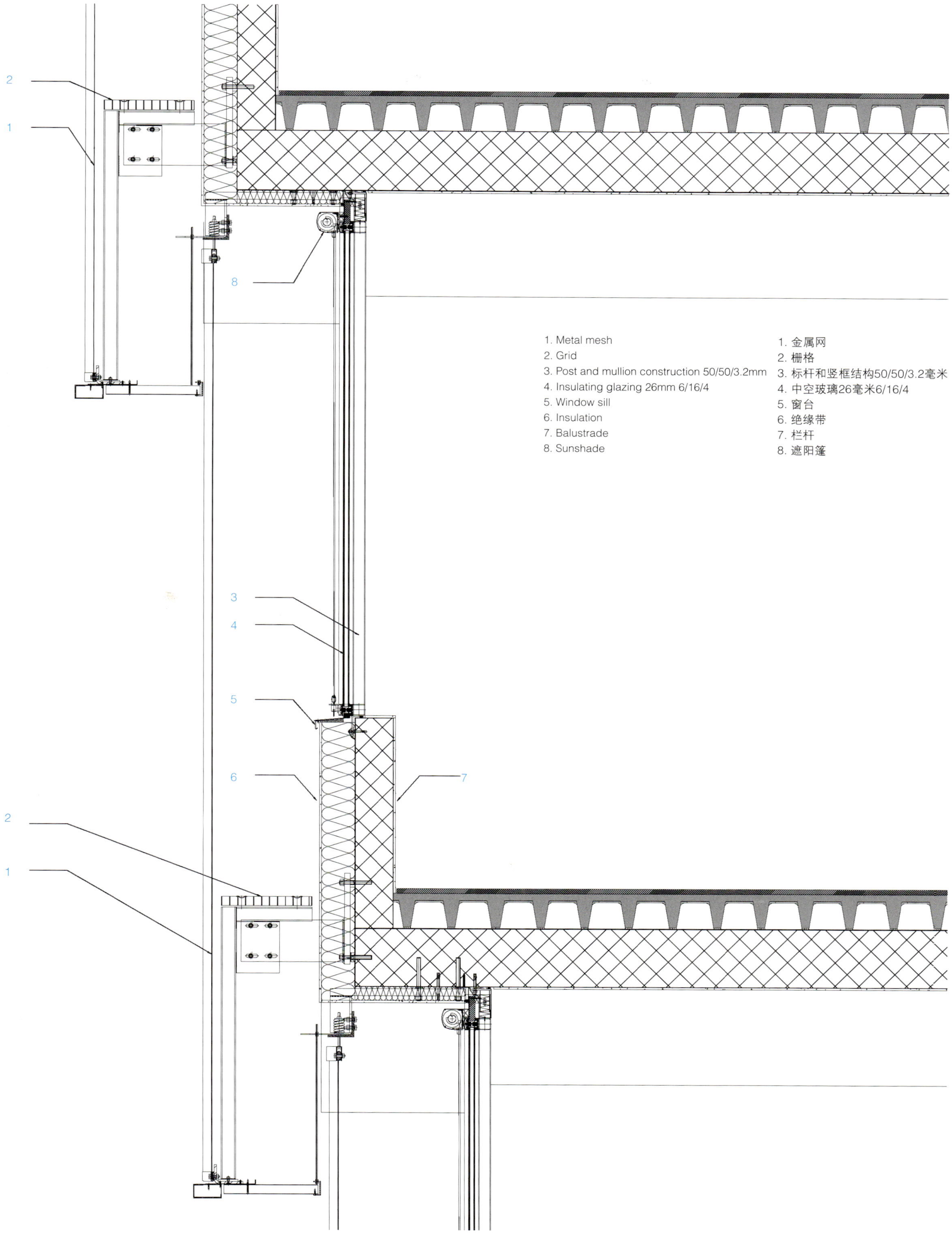

1. Metal mesh
2. Grid
3. Post and mullion construction 50/50/3.2mm
4. Insulating glazing 26mm 6/16/4
5. Window sill
6. Insulation
7. Balustrade
8. Sunshade

1. 金属网
2. 栅格
3. 标杆和竖框结构50/50/3.2毫米
4. 中空玻璃26毫米6/16/4
5. 窗台
6. 绝缘带
7. 栏杆
8. 遮阳篷

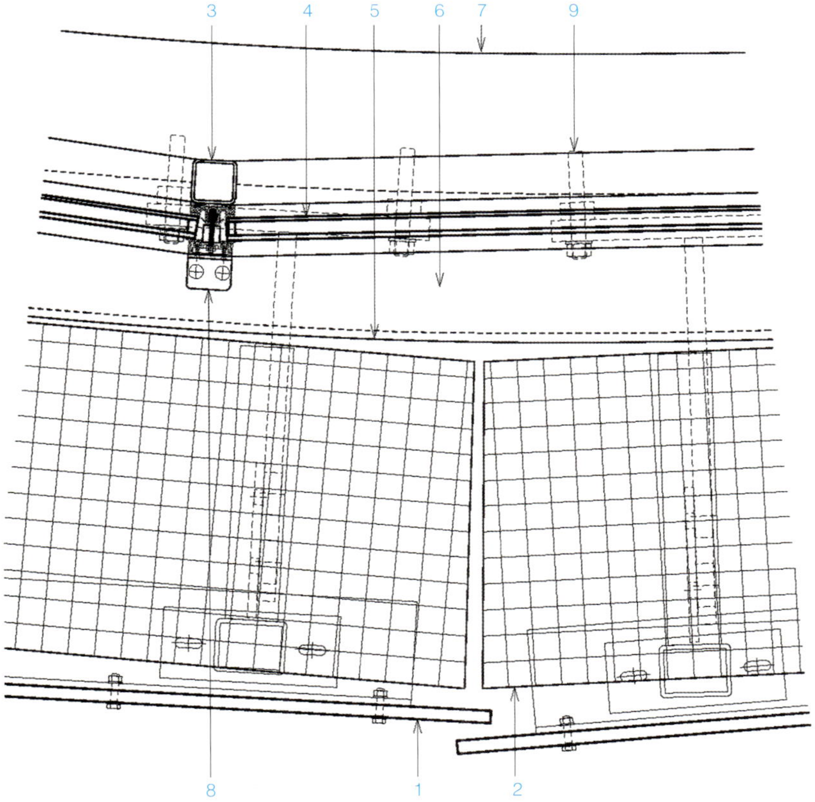

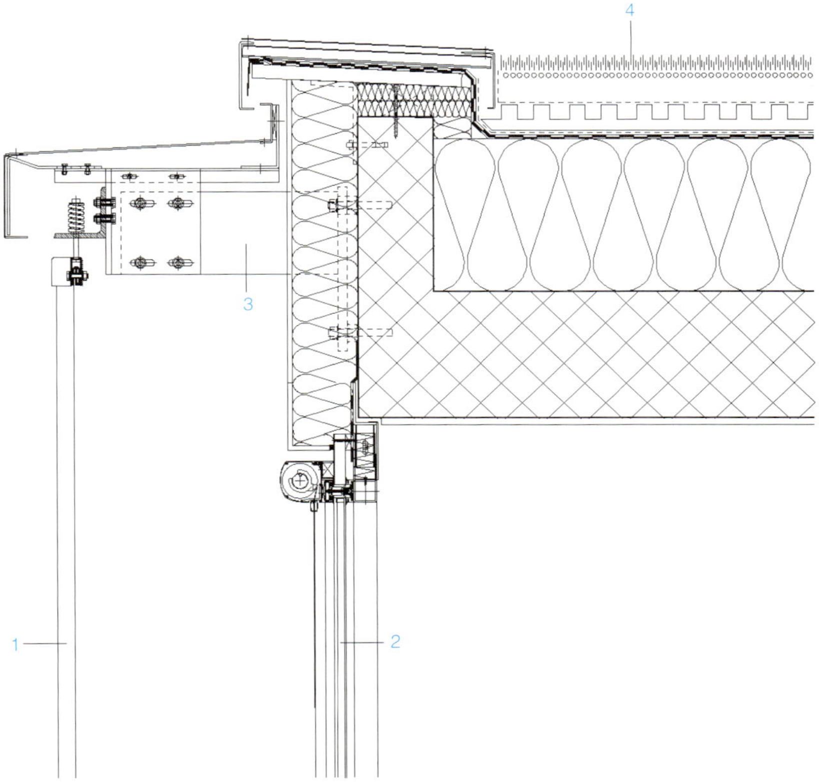

1. Metal mesh
2. Grid
3. Post and mullion construction 50/50/3.2mm
4. Insulation glazing 26mm 6/16/4
5. Window sill
6. Insulation
7. Balustrade
8. Wire cable guides
9. Fixation cantilever

1. 金属网
2. 栅格
3. 标杆和竖框50/50/3.2毫米
4. 绝缘玻璃26毫米6/16/4
5. 窗台
6. 绝缘带
7. 栏杆
8. 电缆导管
9. 固定悬臂

1. Metal mesh
2. Insulating glazing 26mm 6/16/4
3. Cantilever
4. Roof build up:
   Extensive roof greening
   Plant growing substrate 80mm
   Filter fleece 105grams/m^2
   Water storage, drainage layer 75mm
   Root-resistant covering 200grams/m^2
   Bitumen roof sheeting, double layer
   Sloping insulation

1. 金属网
2. 中空玻璃26毫米6/16/4
3. 悬臂
4. 屋顶构成：
   大量的屋顶绿化
   80毫米的植物种植底层
   105克/平方米过滤绒
   5毫米蓄水，排水层
   抵抗性覆层200克/平方米
   双层沥青屋顶板
   绝缘坡

Perimeter Institute for Research in Theoretical Physics

加拿大理论物理研究所

Location: Ontario, Canada
Designers: Saucier+Perrotte Architects
Photographer: Marc Cramer
Completion date: 2007

项目地点：加拿大，安大略省
设计师：Saucier + Perrotte 建筑师事务所
摄影师：马克·克拉默
竣工时间：2007年

Award name:
2001 Architecture Magazine, PA Award, New York
2005 Quebec Association of Architects, Award of Excellence (Institutional)
2005 Ontario Association of Architects, Award of Excellence (Institutional)
2005 Building Magazine "Out of the Box Award"
2006 Governor General's Medal
2006 ArchiZinc Award, Paris
2008 International Architecture Award
2008 Presented by the Chicago Athenaeum and the European Centre for Architecture, Art Design and Urban Studies

获奖名称：
2001年纽约建筑杂志PA奖
2005年魁北克建筑协会优秀建筑奖
2005年安大略省建筑协会优秀建筑奖
2005年建筑杂志最佳创作奖
2006年总督勋章
2006年巴黎ArchiZinc奖
2008年国际建筑奖
2008年芝加哥雅典娜神庙与欧洲建筑、艺术设计与城市研究中心

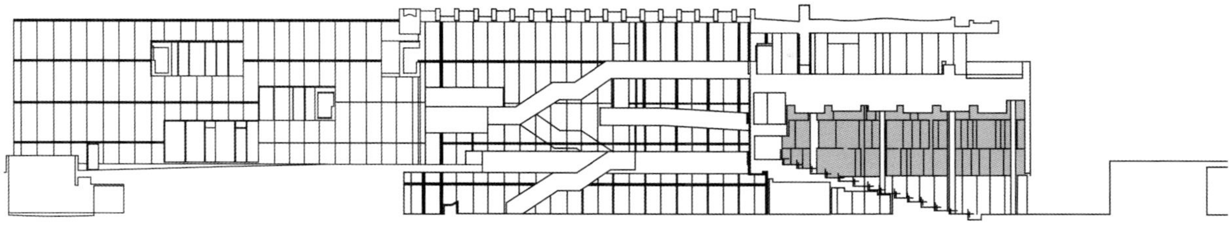

Riding the controversial line between public and private space, this research institute attempts to subvert the usual hard thresholds established by private enterprise in the public realm. The site is on the Shore of Silver Lake, at the northern edge of Waterloo's downtown core and the southern edge of the city's central park. Adjacent to the primary pedestrian access between the university campus and the city centre, the site is an urban wilderness between clearly defined worlds.
The design takes inspiration from the wide-ranging, hard to define concepts that make up the subject matter of theoretical physics, at once micro- and macro-cosmic, rich in information and of indeterminate form and substance. Between city and park, the Perimeter Institute expands and inhabits the improbable space of the line separating the two. The building defines the secure zones of the Institute's facilities within a series of parallel glass walls, embedded in an erupting ground plane that reveals a large reflecting pool. The north façade, facing the park across this pool, reveals the Institute as an organism, a microcosm of discrete elements. The south façade, facing the city across train tracks and the city's main arterial road, presents the Institute as a unified but transforming entity, of enigmatic scale and content. Entry to the Institute is possible from both the north, along the reflecting pool, and the south, under the new ground plane.
The interior of the Institute is organised around two central spaces, the main hall on the ground floor and the garden on the first. Spaces for administration, meeting and seminar rooms, leisure and fitness spaces, and a multipurpose theatre for symposia and public presentations, have direct access to the main hall. The circulation corridors running east-west are positioned between the opalescent glass planes, which are occasionally punctured and shifted to reveal views across the interior space of the hall. Vertical circulation climbs these walls, tendrils of ground that run from the garden through the building. The garden - nature emerging from the vacuum - is crossed by three bridges that puncture all the planes, as well as the north and south façades. The bridges provide quick access to information, facilities and research colleagues. These conduits, which formally bind together the Institute, are routes crossing the improbable space between theoretical physics and everyday life.

研究所建筑究竟属于公共还是私人空间的问题一直存在着争议，而该设计企图颠覆公共空间领域私人机构的传统做法。建筑坐落于“银湖”湖畔，位于铁卢市中心的北部，城市中心公园的南端。临近大学校园和城市中心之间的主要人行道。建筑所在地是都市中一块没有明确界定的区域。
该设计的灵感源自构成理论物理主题、微观和宏观宇宙、不确定形态和内容的一系列不明确概念。该研究所位于城市和公园之间的分界线上。建筑周边的水池围合了双层玻璃墙内的设施，以此对地面的安全范围进行有效界定。北侧立面，隔着水池与公园相对，向人们宣告该研究所是一个有机体、一个由零散元素构成的微型宇宙。南部立面隔着铁轨和主要干线道路面向整个城市，体现出建筑统一而深邃的变幻风格。建筑的南侧和北侧均设置了入口，北入口沿水池而设，南入口则位于新地面层的下方。
研究所的内部围绕着两个中心空间——一楼的大厅和二楼的花园展开布局。管理空间、会议室与研讨室、休闲与健身中心、一个专门进行专题讨论及公众演讲的多功能剧院均与大厅相通。东西向的走廊两边设置的乳白色玻璃板，隐约呈现出大厅的内部风景。垂直交通带从地面升起，沿着玻璃墙面向上攀登，穿过花园，贯穿整个建筑。花园在贯穿所有平面和南北立面的三座桥梁的簇拥下，仿佛从空间中突然“浮现”。桥梁的设计为通往问讯处、设施和研究中心的行人提供了帮助。这些交通纽带从形态上将研究所结合为一个完美的整体，而真正实现了理论物理和日常生活之间的跨越。

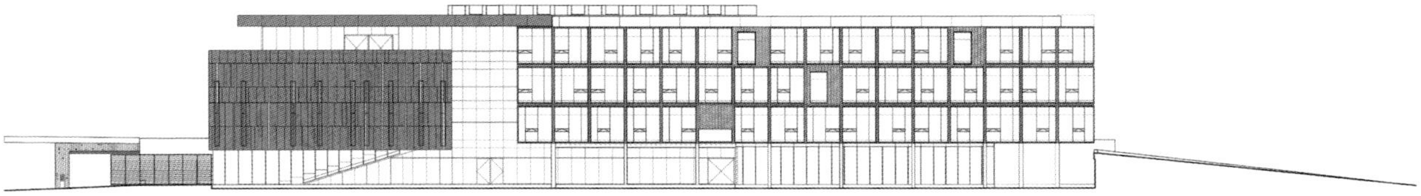

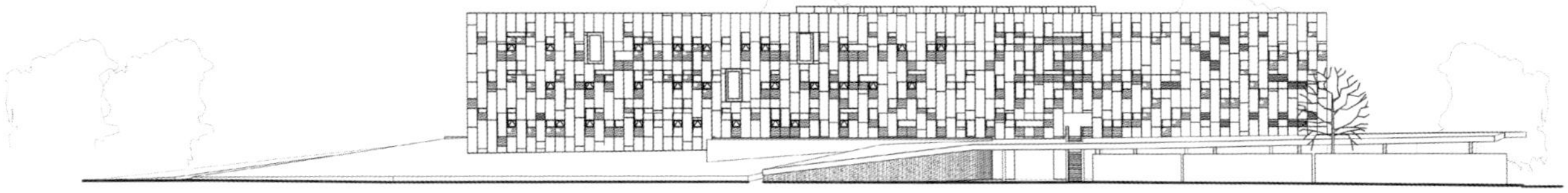

INSTITUTE FOR THEORETICAL PHYSICS
EXIT

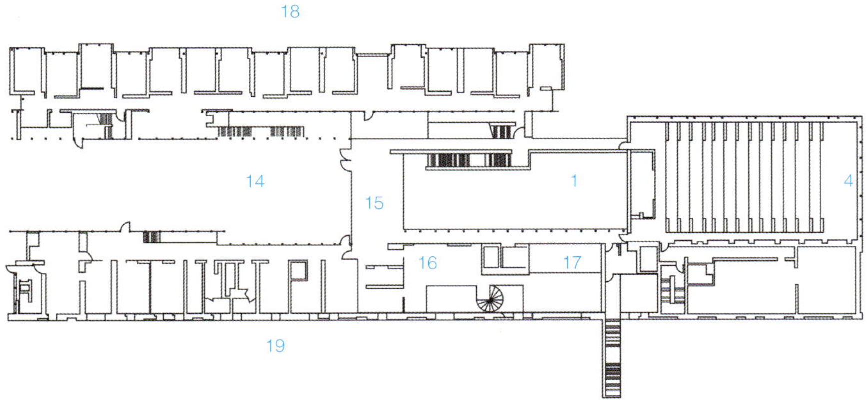

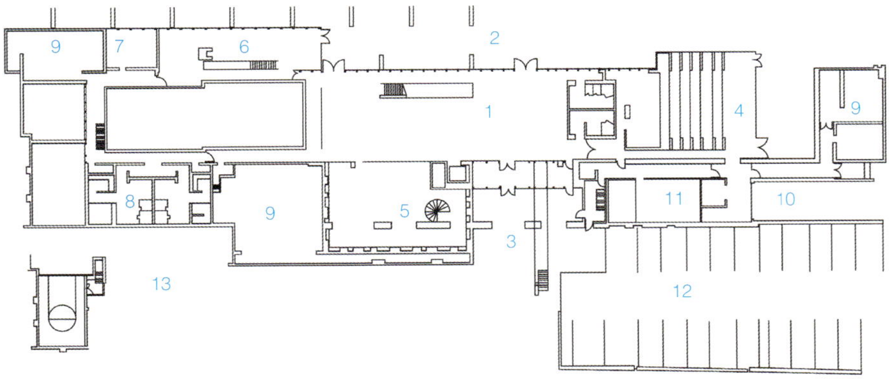

1. Main hall
2. North entrance
3. South entrance
4. Lecture theatre
5. Library
6. Lounge
7. Gym
8. Changing rooms
9. Mechanical room
10. Loading dock
11. Storage
12. Parking
13. Squash (basement)
14. Garden
15. Mezzanine (hall)
16. Mezzanine of the library
17. Reading room
18. Researcher's offices
19. Adminstrative offices

1. 大厅
2. 北入口
3. 南入口
4. 报告厅
5. 图书馆
6. 休息室
7. 健身房
8. 更衣室
9. 机械设备间
10. 装卸区
11. 存储区
12. 停车场
13. 壁球室（地下室）
14. 花园
15. 夹层（礼堂）
16. 图书馆夹层
17. 阅读室
18. 研究人员办公室
19. 行政办公室

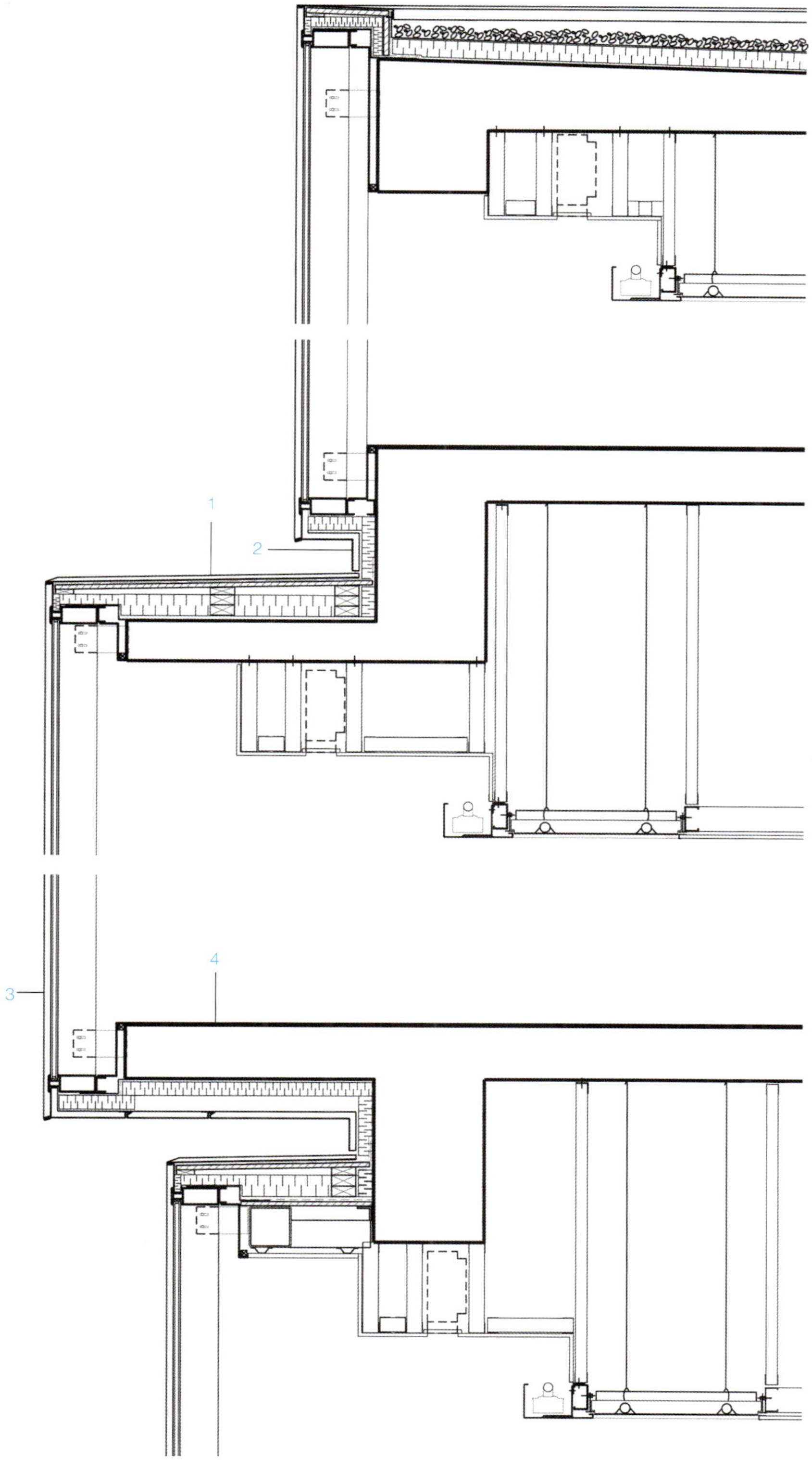

1. Roof composition
   Zinc sheet coping
   Water shield
   16mm treated plywood
   Variable air space
   40mm rigid insulation
   Reinforced membrane
   Site cast concrete
2. Wall composition
   25mm interlock zinc panel
   100mm adjustable bars
   10mm air space
   50mm polyurethane insulation
   Blueskin membrane
   Site cast concrete
3. Curtain wall
4. Floor composition
   Site cast concrete
   Blueskin membrane
   75mm polyurethane insulation
   25mm air space
   100mm adjustable angles
   Zinc interlock panels

1. 屋顶构成：
   顶部锌板
   防水罩
   16毫米特种胶合板
   通风空间
   40毫米的硬性绝缘带
   加固膜
   现浇混凝土
2. 墙体构成：
   25毫米的连锁锌面板
   100毫米的可调式钢筋
   10毫米的通风空间
   50毫米聚氨酯绝缘带
   蓝皮薄膜
   现浇混凝土
3. 幕墙
4. 楼板构成：
   现浇混凝土
   蓝皮薄膜
   75毫米聚氨酯绝缘板
   25毫米通风空间
   100毫米可调角度
   镀锌连锁板

Rotermanni Vana Jahuladu ja Uus Jahuladu

Rotermann Jahuladu办公楼

Location: Tallinn, Estonia
Designers: HG Arhitektuur
Photographers: Yoko Azukawa, Hanno Grossschmidt, Tomomi Hayashi/ HG Arhitektuur
Completion date: 2009

项目地点：爱沙尼亚，塔林
设计师：HG 建筑师事务所
摄影师：东川阳子，汉诺・克罗斯密特，林知充/HG建筑师事务所
竣工时间：2009年

Award name:
Architectural Prize 2009 from the Estonian Cultural Endowment
Best Building in Historical Environment 2008 from the Culture Department of City of Tallinn

获奖名称：
2009年爱沙尼亚文化基金会建筑奖
2008年塔林市文化部历史环境最佳建筑奖

The 15 century-old Rotermann Quarter, a former industrial area for food production, is located between the Tallinn's old town and the port, where stands still historically-valuable limestone buildings under heritage protection. The approved detail planning calls for adoptive re-use of existing buildings as well as insertion of new volumes amongst in order to create “live-work” and pedestrian-friendly environment right at the city centre, supported by 400 parking lots underground. On-going redevelopment takes place between the existing historical limestone buildings, including the Old Flour Storage from 1904. The New Flour Storage was aimed to form a plaza as a new focal point of the quarter. The project consists of three volumes; the Old Flour Storage with two additional stories, the New Flour Storage and the Atrium connecting the two. Ground floor is for retail and all upstairs are for offices.
The approach was to relate and strengthen the character of historical quarter through finding and adopting the character of the surroundings. For façade articulation the designers have abstracted proportion of wall versus window openings as a character of old industrial buildings. For main façade material Cor-Ten steel was chosen for its property fitting to the existing surroundings of rough surfaces; limestone walls, brick lintels and rusted steel details. It pays homage to the area's industrial past.
The New Flour Storage is to be a “Communication Wall” between offices and the plaza. The windows consist of three sizes; 75centimetres x 75centimetres, 2metres x 2metres, larger ones over floor height. The small ones are to frame the views as well as to bring fresh air, the middle ones are to relate to the human scale, and the biggest is to provide the panorama view of the plaza and the Old Town. The cantilevered bigger windows breaks the homogeneous rectangular outline of the building and offers special place among office space for meeting room or communal area with panorama views.

拥有150年历史的塔林市Rotermann区，其前身是食品生产工业所在地，介于塔林老城区和港口之间，因石灰岩历史保护性建筑而著称。该项目的设计要求在原有建筑的基础上增设新的体量，旨在为城市中心打造一个宜居的办公环境，并为行人提供优雅的散步场所。同时，地下还设置了400个停车位。此外，设计师还要对原有的历史石灰石建筑（包括建于1904年的老面粉储藏库）进行改建，并在打造新面粉储藏库的同时，构建一个可作为该地区焦点的广场。该项目由三部分体量构成，它们分别是增设两层空间后的老面粉储藏库、新面粉储藏库以及将二者进行有机衔接的中庭。一层空间中设有零售店，楼上空间则专属为办公区。
设计手法是建立并加强建筑与周围环境之间的联系，并突出历史保护区的特点。对于建筑的立面，设计师通过将抽象的墙体比例与开窗结构之间的对比有力地彰显出老工业建筑的特征。建筑的主立面采用科尔坦耐大气腐蚀高强度钢材料，与周围粗糙的表面——石灰石墙壁、砖楣、生锈钢零件等交相辉映，向人们倾诉着该地区古老的工业传奇。

新面粉储藏库在办公空间和广场之间扮演着“交流墙”的角色。窗口的尺寸分别为75x 75厘米、2x 2米和全高。尺寸较小的窗户起到打造景观、加强室内通风的作用；2x 2米的尺寸在其他建筑中较为普遍；全高的玻璃窗则能够帮助室内的人们将广场和老城区的全景尽收眼底。悬垂的大型玻璃窗打破了建筑整齐的矩形轮廓，巧妙地在办公空间中构造出专为会议室和公共区域空间设置的特殊区域，开阔的空间，欣赏景观的角度极佳。

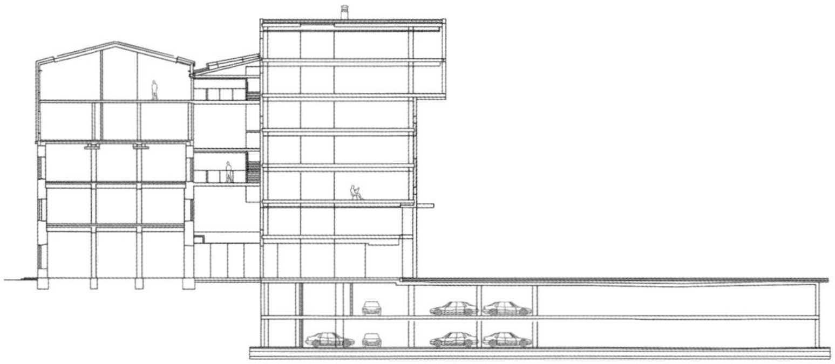

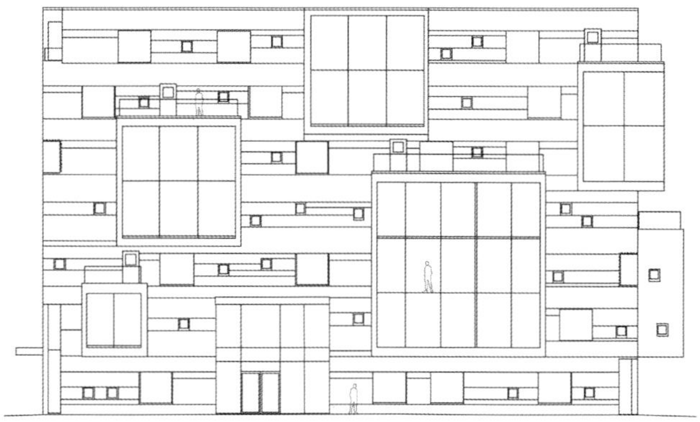

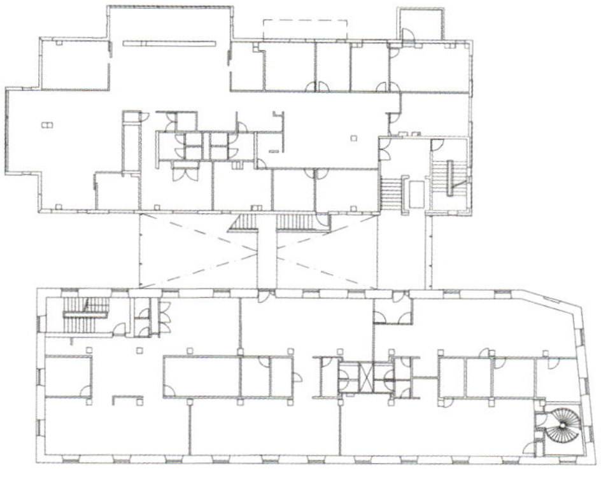

rotermann

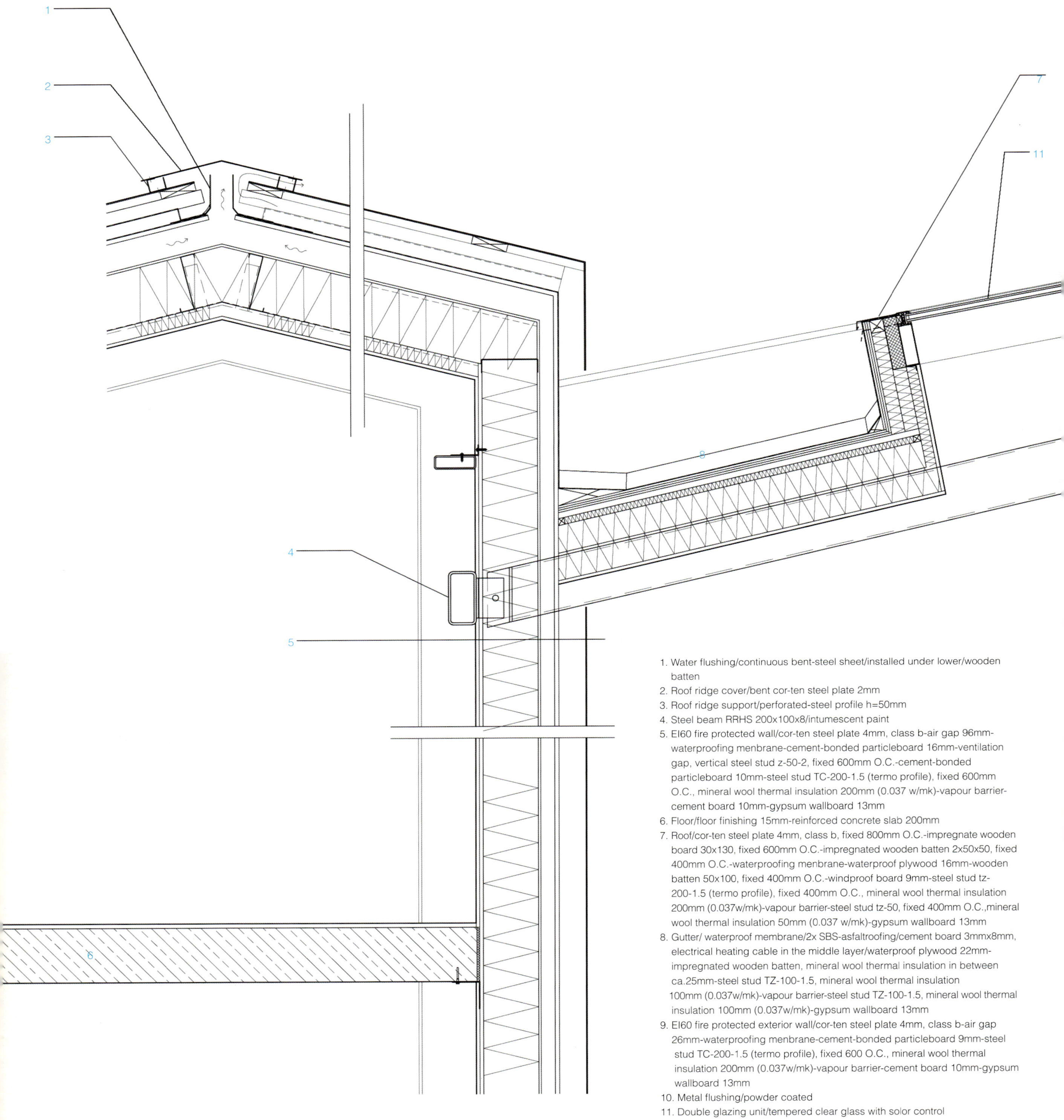

1. Water flushing/continuous bent-steel sheet/installed under lower/wooden batten
2. Roof ridge cover/bent cor-ten steel plate 2mm
3. Roof ridge support/perforated-steel profile h=50mm
4. Steel beam RRHS 200x100x8/intumescent paint
5. EI60 fire protected wall/cor-ten steel plate 4mm, class b-air gap 96mm-waterproofing menbrane-cement-bonded particleboard 16mm-ventilation gap, vertical steel stud z-50-2, fixed 600mm O.C.-cement-bonded particleboard 10mm-steel stud TC-200-1.5 (termo profile), fixed 600mm O.C., mineral wool thermal insulation 200mm (0.037 w/mk)-vapour barrier-cement board 10mm-gypsum wallboard 13mm
6. Floor/floor finishing 15mm-reinforced concrete slab 200mm
7. Roof/cor-ten steel plate 4mm, class b, fixed 800mm O.C.-impregnate wooden board 30x130, fixed 600mm O.C.-impregnated wooden batten 2x50x50, fixed 400mm O.C.-waterproofing menbrane-waterproof plywood 16mm-wooden batten 50x100, fixed 400mm O.C.-windproof board 9mm-steel stud tz-200-1.5 (termo profile), fixed 400mm O.C., mineral wool thermal insulation 200mm (0.037w/mk)-vapour barrier-steel stud tz-50, fixed 400mm O.C.,mineral wool thermal insulation 50mm (0.037 w/mk)-gypsum wallboard 13mm
8. Gutter/ waterproof membrane/2x SBS-asfaltroofing/cement board 3mmx8mm, electrical heating cable in the middle layer/waterproof plywood 22mm-impregnated wooden batten, mineral wool thermal insulation in between ca.25mm-steel stud TZ-100-1.5, mineral wool thermal insulation 100mm (0.037w/mk)-vapour barrier-steel stud TZ-100-1.5, mineral wool thermal insulation 100mm (0.037w/mk)-gypsum wallboard 13mm
9. EI60 fire protected exterior wall/cor-ten steel plate 4mm, class b-air gap 26mm-waterproofing menbrane-cement-bonded particleboard 9mm-steel stud TC-200-1.5 (termo profile), fixed 600 O.C., mineral wool thermal insulation 200mm (0.037w/mk)-vapour barrier-cement board 10mm-gypsum wallboard 13mm
10. Metal flushing/powder coated
11. Double glazing unit/tempered clear glass with solor control
12. Steel beam RRHS 200x100x8/intumescent paint
13. Fire curtain unit

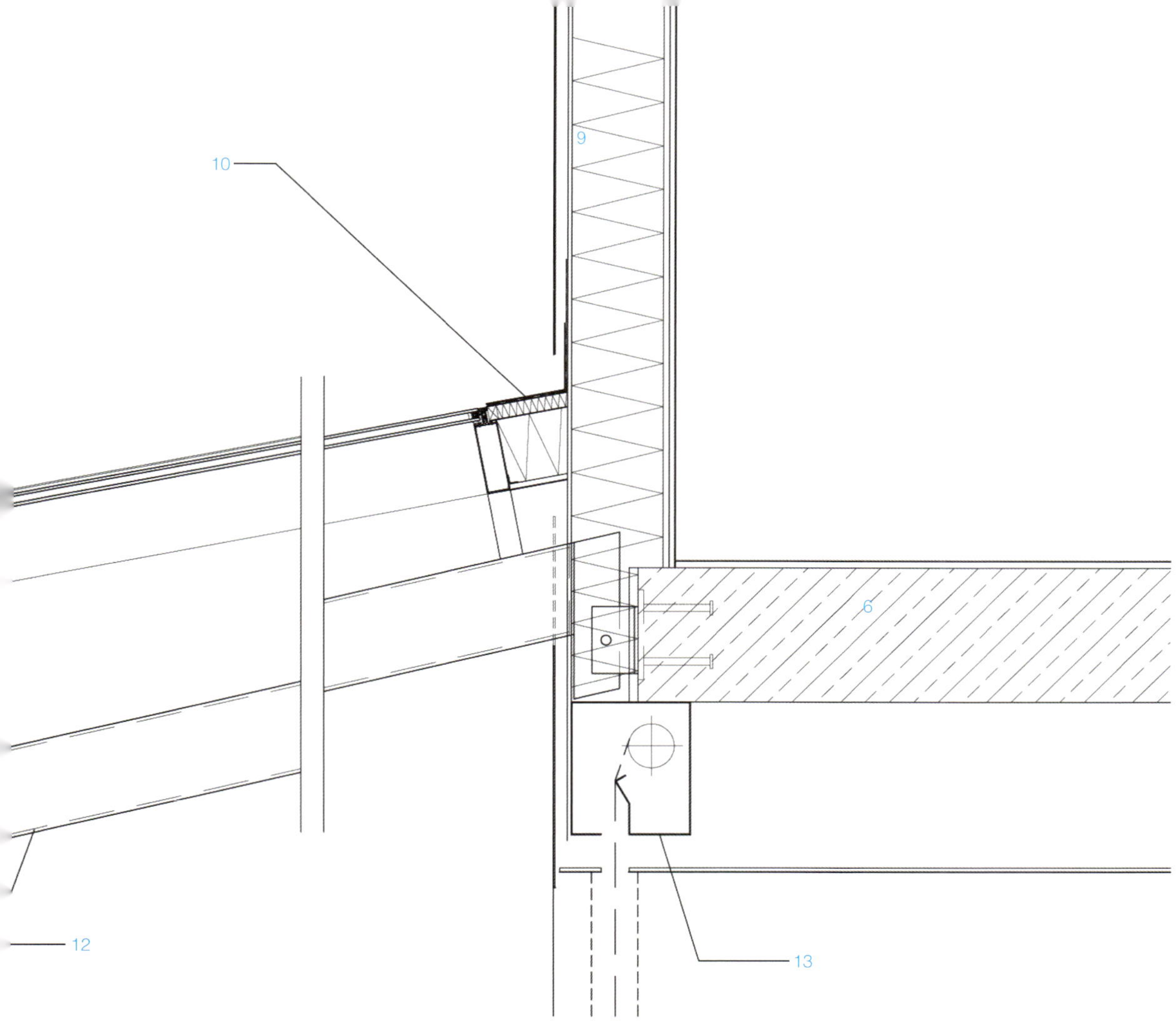

1. 冲水/ 连续弯钢板/低层安装/木板条
2. 屋脊盖/科尔坦耐大气腐蚀高强度弯曲钢板，2毫米
3. 屋脊支撑/穿孔钢材剖面50毫米
4. 钢梁RRHS 200x100x8/防火漆
5. EI60防火保护墙；科尔坦耐大气腐蚀高强度钢板4毫米，B级；96毫米空气间隙；防水膜；水泥刨花板16毫米；通风间隙，垂直钢螺柱z-50-2，固定的600毫米中心间距；水泥结合芯板材10毫米；钢螺柱TC-200-1.5，固定的600毫米中心间距；矿物棉绝缘层200毫米；蒸汽屏蔽；水泥板10毫米；石膏墙板13毫米
6. 地板/地面层15毫米，200毫米钢筋混凝土板
7. 屋顶；科尔坦耐大气腐蚀高强度钢板厚度为4毫米，B级，固定的800毫米中心间距；浸染木板30 × 130毫米，固定的600毫米中心间距；浸染木板条 2 × 50 × 5 0 毫米，固定的400毫米中心间距；防水膜；16毫米耐水胶合板；木板条50x100毫米，固定的400毫米中心间距；防风板 9 毫米；钢螺柱TZ-200-1.5，固定的400毫米中心间距；矿物棉绝缘层50毫米，石膏墙板13毫米
8. 排水沟；防水薄膜/2x SBS沥青屋顶;水泥板3x8毫米；中层设有加热电缆；22毫米耐水胶合板；浸染木板条；矿物棉绝缘保温层约25毫米；钢螺柱TZ-100-1.5；矿物棉绝缘保温层约100毫米；蒸汽屏蔽；钢螺柱TZ-100-1.5；矿物棉绝缘保温层约100毫米；石膏墙板13毫米
9. EI60防火护墙板；科尔坦耐大气腐蚀高强度钢板4毫米，B级；26毫米空气间隙；防水膜；水泥刨花板9毫米；钢螺柱TC-200-1.5，固定的600毫米中心间距；矿物棉绝缘层200毫米；蒸汽屏蔽；水泥板10毫米；石膏墙板13毫米
10. 金属冲洗/粉末涂层
11. 双层玻璃装置/热处理耐温透明玻璃
12. 钢梁RRHS 200x100x8/防火漆
13. 防火幕

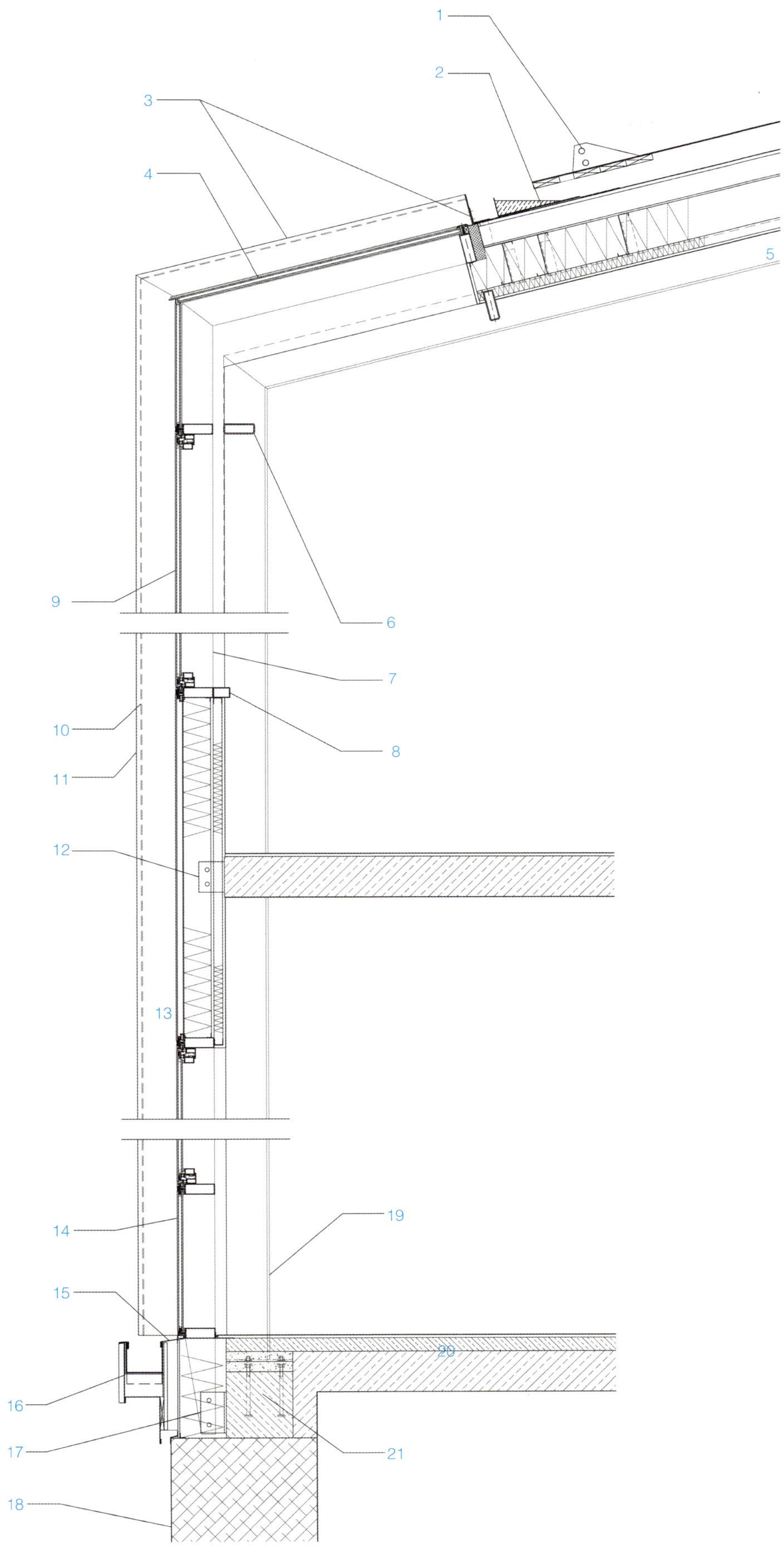

1. Snow stopper/steel tube/powder coated
2. Gutter/bent steel sheet h=135/ powder coated
3. Cover plate/al-sheet thickness=4mm, h=150/powder coated
4. Double glazing unit/tempered clear glass with solar control
5. Roof/cor-ten steel plate 4mm, class b, fixed 800mm O.C.-impregnated wooden board 30x130, fixed 600mm O.C.-impregnated wooden batten 2x50x50, fixed 400mm O.C.-waterproofing menbrane-waterproof plywood 16mm-wooden batten 50x100, fixed 400mm O.C.-windproof board 9mm-steel stud TZ-200-1.5 (termo profile), fixed 400mm O.C.-mineral wool thermal insulation 200mm (0.037w/mk)-vapour barrier-steel stud TZ-50, fixed 400mm O.C.-mineral thermal insulation 50mm (0.037w/mk)-gypsum wallboard 13mm
6. Steel frame RRHS 150x50x5.0/intumescent paint
7. Vert. window frame/al-profile 50x140mm/powder coated
8. Window sill/al-profile 50x80mm/powder coated
9. Double glazing unit/tempered clear glass
10. Cor-ten steel plate behind thickness=4mm
11. Cover plate/al-plate thickness=4mm, width=185mm
12. Vert.window frame fixed to the edge of slab
13. Spandrel insulated panel/tinted glass dark bronze 6mm-cement board 9mm-mineral wool thermal insulation-gypsum wall board 13mm-steel stud c-50, mineral wool thermal insulation-sypsum wall board 13mm
14. Double glazing unit/tempered clear glass
15. Metal flushing/powder coated
16. Gutter/bent steel sheet/powder coated
17. Vert.window frame fixed to the edge of slab
18. Exiting limestone wall
19. Steel frame hea220/intumescent paint
20. Floor/floor finishing 15mm-reinforced concrete topping 60mm-80mm-existing prefabricated concrete slab
21. Continuous R/C footing on top of ext. limestone wall

1. 雪塞/钢管/粉末涂层
2. 排水沟/弯钢板　高= 135毫米 /粉末涂层
3. 盖板/铝板T = 4毫米，高= 150毫米/粉末涂层
4. 双层玻璃装置/热处理隔热透明玻璃
5. 屋顶；科尔坦耐大气腐蚀高强度钢板厚度为4毫米，B级，固定的800毫米中心间距；浸染木板30 x 130毫米，固定的600毫米中心间距；浸染木板条2 x 50 x 50毫米，固定的400毫米中心间距；防水膜；16毫米耐水胶合板；木板条50x100毫米，固定的400毫米中心间距；防风板 9 毫米；钢螺柱TZ-200-1.5，固定的400毫米中心间距；矿物棉绝缘层50毫米，石膏墙板13毫米
6. 钢架RRHS150x50x5.0毫米/防火漆
7. 垂直窗框/铝合金50x140毫米/粉末涂层
8. 窗台板/铝合金50x80毫米/粉末涂层
9. 双层玻璃装置/热处理透明玻璃
10. 科尔坦耐大气腐蚀高强度钢板，厚度为4毫米
11. 盖板/铝板　厚度= 4毫米，宽= 185毫米
12. 窗框固定在楼板边缘
13. 拱肩绝缘板/茶色玻璃6毫米水泥板，9毫米岩棉保温石膏墙板，13毫米钢螺栓，矿棉保温绝缘石墙板13毫米
14. 双层玻璃装置/热处理透明玻璃
15. 耐冲金属/粉末涂层
16. 槽/弯钢板/粉末涂层
17. 窗框固定在楼板边缘
18. 出口石灰石墙
19. 防水钢架220/防火漆
20. 地板/地板层15毫米，钢筋混凝土平顶60毫米，80毫米预制混凝土板
21. 连续的R/C底脚位于灰岩墙延伸部分的上方

pinnad
innaga!
büroopinnad

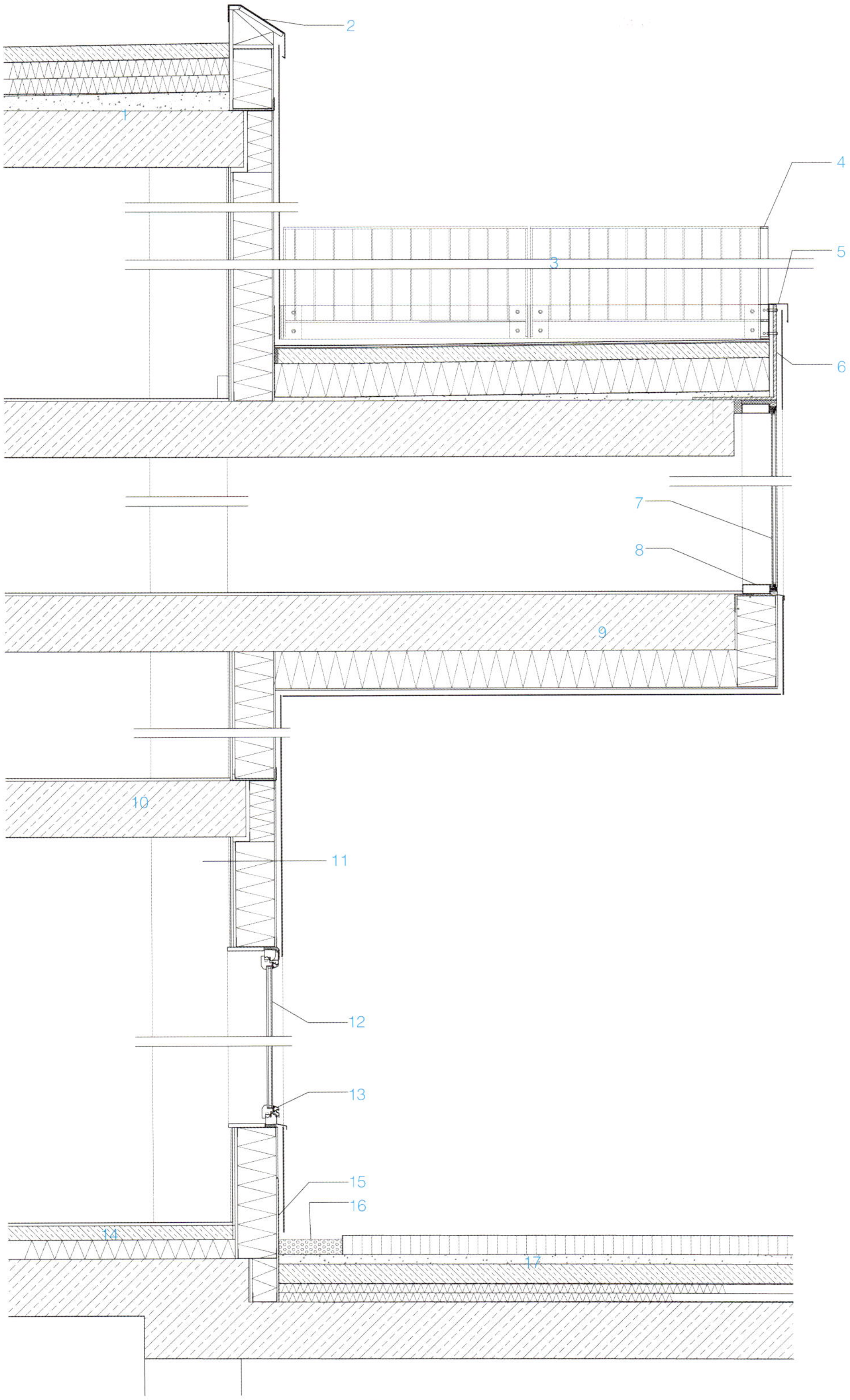

1. Roof/reinforced concrete slab 80mm-filtering membrane-thermal insulation xps 180mm (0.037w/mk)-waterproofing membrane-autoclaved aerated concrete (AAC) filling-reinforced concrete slab 200mm
2. Parapet coping/steel sheet/powder coated
3. Terrace/ceramic tile-cement grout-reinforced concrete slab 80mm-filtering membrane-xps thermal insulation 180mm (0.037w/mk)-waterproofing membrane-autoclaved aerated concrete (AAC) filling-reinforced concrete slab 200mm
4. Railing/stainless steel flat bar
5. Parapet coping/steel sheet/powder coated
6. Steel support for railing
7. Double glazing unit/tempered clear glass with solor control
8. HOR.window frame/AL-profile 50x140mm/powder coated
9. Cantilevered floor/floor finishing 15mm-reinforced concrete slab 300mm-steel stud TC-225-1.2 fixed 600mm O.C., thermal insulation 180mm xps (0.037w/mk)-windproof board 9mm-air gap 26mm-cor-ten steel plate 4mm, classb
10. Floor/floor finishing 15mm-reinforced concrete slab 200mm
11. Exterior wall/cor-ten steel plate 4mm, class b-air gap 26mm-windproof board 9mm-steel stud TC-200-1.5 (termo profile), fixed 600mm O.C., mineral wool thermal insulation 200mm (0.037w/mk)-vapour barrier-2x gypsum wallboard 13mm
12. Double glazing unit/tempered clear glass
13. Aluminium clad wooden window
14. Floor/floor finishing 15mm-reinforced concrete slab 75mm-filter membrane-thermal insulation eps 100mm-reinforced concrete slab 300mm
15. Cement board 10mm
16. Gravel dia=20mm
17. Plaza/granite cube-sand 50mm-reinforced concrete slab 100mm-filtering menbrane-thermal insulation 180mm xps (0.037w/mk)-waterproofing membrane-reinforced concrete slab 300mm

1. 屋顶/钢筋混凝土板80毫米–过滤膜保温绝缘XPS 180毫米（0.037w/mk）– 防水卷材高压蒸汽养护的加气混凝土（AAC）灌装钢筋混凝土板200毫米
2. 护墙顶部/钢板/粉末涂层
3. 阳台/瓷砖水泥灌浆钢筋混凝土板，80毫米过滤膜XPS保温绝缘，180毫米（0.037w/mk）防水卷材，蒸汽加压混凝土（AAC）灌装钢筋混凝土板200毫米
4. 栏杆/不锈钢扁钢
5. 护墙顶端/钢板/粉末涂层
6. 钢支架围栏
7. 双层玻璃装置/热处理耐温透明玻璃
8. 水平窗框/铝剖面50x140毫米/粉末涂层
9. 悬臂式地板/地板层15毫米，钢筋混凝土板300毫米，钢螺栓TC–225–1.2，固定的600毫米中心间距，保温绝缘板 180毫米（0.037w/mk），防风板9毫米，空气间隙26毫米，科尔坦耐大气腐蚀高强度钢板 4毫米，B级
10. 地板/地板装修15毫米，200毫米钢筋混凝土板
11. 外墙/科尔坦耐大气腐蚀高强度钢板 4毫米，B级，空气间隙26毫米，防风板9毫米，钢螺栓TC–200–1.5（保温侧面），固定的600毫米中心间距，矿棉保温板200毫米（0.037w/mk），蒸汽屏蔽，两个13毫米厚的石膏墙板
12. 双层玻璃装置/热处理耐温透明玻璃
13. 铝包木窗
14. 地板/地板层15毫米，75毫米钢筋混凝土板，过滤膜保温板100毫米，300毫米钢筋混凝土板
15. 10毫米的水泥板
16. 碎石直径20毫米
17. 广场/花岗岩立方砂50毫米，100毫米钢筋混凝土板，180毫米热绝缘过滤细胞膜XPS（0.037w/mk），防水卷材– 钢筋混凝土板300毫米

Steckelhörn 11 - Office Complex

Steckelhörn 11综合办公大楼

Location: Hamburg, Germany
Designers: J. MAYER H. Architects
Project team: Juergen Mayer H., Hans Schneider, Wilko Hoffmann, Marcus Blum
Photographer: David Franck
Completion date: 2009

项目地点：德国，汉堡
设计师：J. MAYER H.建筑师事务所
项目组：尤根・梅尔・H.，汉斯・施耐德，威尔科・霍夫曼，马库斯・百隆
摄影师：大卫・弗兰克
竣工时间：2009年

The project "Steckelhörn 11" is located in the old centre of Hamburg, close to the prominent new "Hafen City" development. It replaces a ruinous building and fills the gap between two historic premises. The triangular-shaped lot stretches across the city block, thus allowing for a narrow façade of about 1.3 metres width facing the harbour and a main elevation of about 26.4 metres oriented towards the Steckelhörn Street.

The vertical design and soft setbacks of the latter pay tribute to the massing of the surrounding structures, as well as to local building-height regulations. Cantilevered elements in the main façade create a series of specific spatial qualities on the inside and outside. The top floors provide additional outside space, offering a spectacular panoramic view over the city of Hamburg.

The particular geometry of the floor plan is the basis for the organisation of the building, which architecturally and programmatically presents itself openly to the Steckelhoern Street while at the same time forming a characteristic landmark when perceived from the historic "Speicherstadt" and new "Hafen City". As the ground level is conceived either as a spacious lobby for the main tenant or a public cafe, the upper floors provide for generous, flexible office space, most of it allowing a view of the "Katharinenkirche" and/or the "Hafen City". The top floors provide additional outside space on balconies/loggias and a roof terrace, offering a spectacular panoramic view over the old and new city of Hamburg.

Steckelhörn 11综合办公大楼位于汉堡的老城中心，毗邻著名的"Hafen 城市"开发区。新建筑在一个破败的建筑基础上进行重建，重建后填补了两个具有历史意义的基地之间的空白。根据周围建筑的结构和高度，该建筑呈三角形设计，1.3 米宽的狭窄的外立面与港口遥遥相对，26.4 米宽的主外立面直面Steckelhörn大街。

垂直状态设计和柔和的墙体缩进与周围建筑完美、和谐地衔接在一起，并与当地建筑高度标准相得益彰。主外立面上的悬臂式结构在建筑内外打造出一系列极富个性的空间。顶层为住户提供了额外的户外空间，在这里，人们可以将壮美的汉堡城市景色尽收眼底。

造型独特的几何形状平面布局基于建筑的结构而设计。从建筑本身和规划程序的角度来看，建筑在面向Steckelhörn大街的同时，为具有历史意义的Speicherstadt 城区和新的"Hafen城区 "之间打造了一个醒目的地标。建筑的一层设有一个宽敞的大厅，上层为宽阔且灵活的办公空间，能够饱览"Katharinenkirche"地区或 "Hafen城区"美景。顶楼在露台/凉廊处增设了一个室外空间以及一个屋顶露台，将壮观的汉堡新老城区美景尽情展现在住户面前。

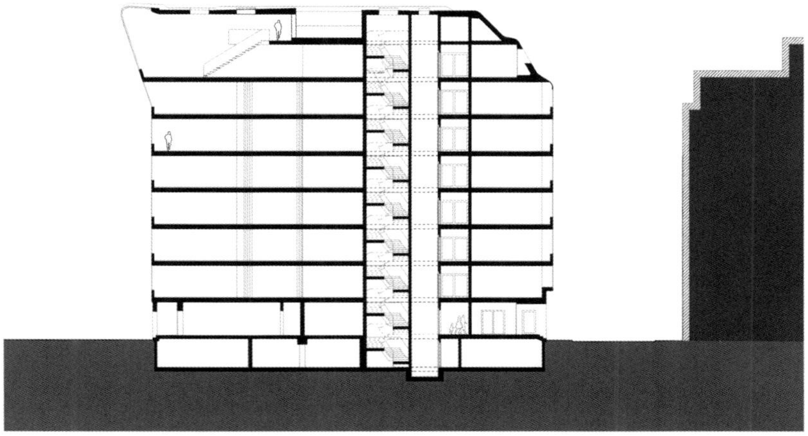

STECKELHORN 11

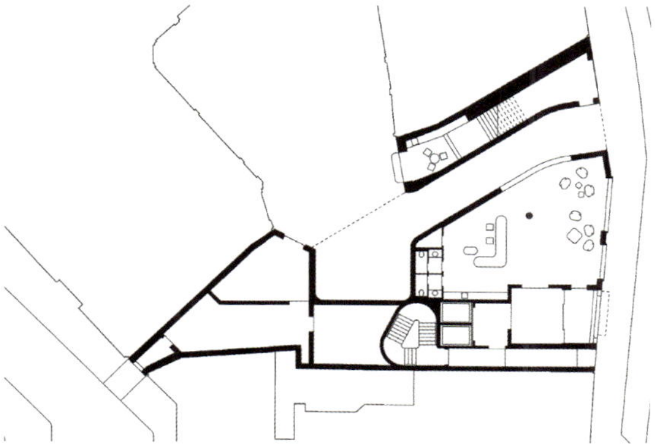

RESTAURANT

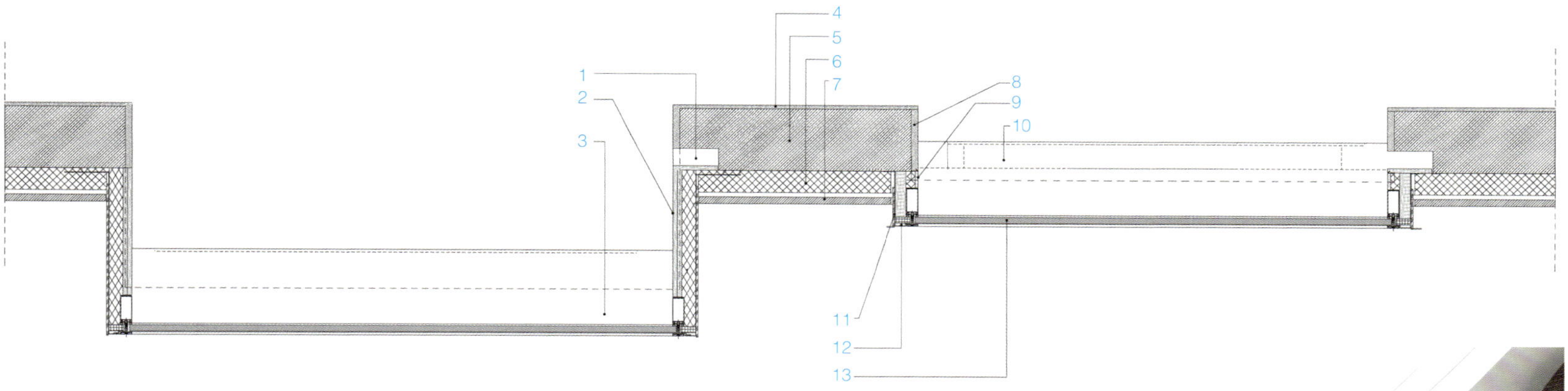

1. Slot for glare shield
2. Dry walls covering
3. Windowsill, aluminium, powder-coated
4. Plaster 1.5cm
5. Steel concrete wall 28cm
6. Heat insulation
7. Non-bearing ceramic elements
8. Ca. 3cm plaster
9. Dry walls covering
10. Konvector
11. Aluminium sheet, 8mm, anodised
12. Aluminium façade, anodised
13. Safety glass

1. 遮光槽
2. 干砌墙墙面涂料
3. 窗台、铝、粉末涂层
4. 石膏1.5厘米
5. 钢筋混凝土墙28厘米
6. 隔热带
7. 非承重陶瓷元件
8. 约为3厘米的石膏
9. 干砌墙墙面涂料
10. 对流式加热器
11. 8毫米厚阳极氧化处理铝板
12. 阳极氧化铝材表面
13. 安全玻璃

Financial and Commercial Department of Voestalpine Stahl GmbH

奥钢联斯塔尔有限公司金融与商业部办公大楼

Location: Linz, Austria
Designers: Dietmar Feichtinger Architectes
Photographers: Josef Pausch
Completion date: 2009

项目地点：奥地利，林茨
设计师：Dietmar Feichtinger建筑师事务所
摄影师：约瑟夫·鲍什工作室
竣工时间：2009年

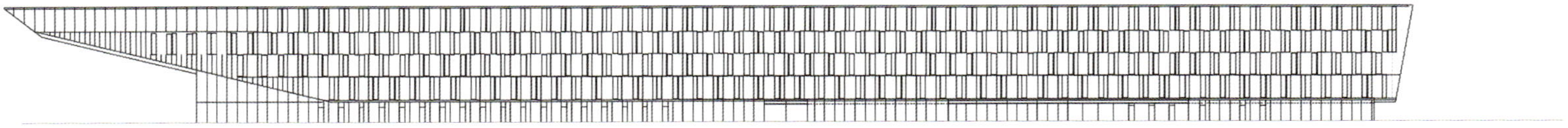

The new sales and financial centre and the existing "Blue Tower" form the entry to the plant of Voestalpine: two complementary structures representing the complementary units of Voestalpine. The horizontal shape of the new building dialogues with the existing vertical tower. The building is seen already from a fardistance. The cantilever of several storeys covering the entrance area of the building is an inviting gesture. It symbolises the ambition of a modern and innovative company. The open access area is a meeting place.

A gently sloping surface leads to the entrances. Located 1m above the existing terrain a new platform connects the buildings. It is reserved for pedestrians only. It is covered by large-size concrete elements - coloured sand-blasted in situ concrete. LED lights are integrated into the surface. The low level of illumination of the open spaces contrasts with the brightly lit entrance of the building.

By its geometry the new sales and financial centre defines a generous open public space covering a one storey parking deck. The spacious garden creates a designed outdoor area for employees and visitors of the company: a sequence of green fields, wooden decks, paved and planted areas structured by patios.

新销售与金融中心与原有的“蓝塔”建筑一起为奥钢联斯塔尔有限公司打造了一个完美的入口：相辅相成的两个结构象征着奥钢联斯塔尔有限公司各部门之间的相互支撑与互动。水平形态的新建筑与原有的垂直大楼实现了空间上的对话。建筑在远处即清晰可见。覆盖建筑入口的几层悬臂结构呈盛情邀请状，象征着该公司现代、创新的运营理念。开放空间可作为一个大型会客区。

一个平缓的斜坡界面向入口处徐徐靠近。高出地面1米的新平台将建筑有机衔接在一起。该平台仅作为行人专用区，以大型混凝土元素 – 彩色喷砂现浇混凝土为主要原料。建筑的表面引用了LED照明设备。低层公共区的照明设备与建筑入口明亮的照明效果形成鲜明的视觉对比。

造型独特的几何型销售与金融中心定义了一个开阔的公共开放空间，并巧妙打造了一层停车区。宽敞的室外花园区为员工和访客打造了一个别致的人工室外休息空间：绵延的绿地、木板平台、露台构建的铺石区和种植区。

The parking spaces are located under this new platform. It is an open, naturally ventilated and naturally lit up space. Planted courts a natural soil connect the level to the outside. Its access road follows the curved rail tracks. This road also leads to the exhibition building called “the world of steel”.

The structure of the building is a combined steel and concrete structure. The columns are in the centre of the building, the side areas are carried by cantilever beams. This configuration offers a large flexibility for office division. The concrete slabs are kept visible. They provide the mass necessary for thermal comfort. An integrated steel truss allows the free cantilever in the front of the building. The façade is largely transparent providing sufficient natural light for the building. Opaque horizontally sliding openings allow natural ventilation. Shading is assured by motorised steel elements made of a golden coloured steel mesh.

Facing the new building the ground floor of the existing building is renewed. A large glass canopy covers the entrance area. Its steel structure is anchored in the structure of the existing building.

位于新平台下方的停车区内开阔、通风性能良好。种植的植物将该地区与外部空间巧妙衔接。其入口与弧形的轨道相通。该路线同时也可通向“钢材世界”展览大楼。

该建筑采用钢筋和混凝土相结合的结构。建筑的中央设置若干梁柱，侧面区域由悬臂梁进行支撑。该结构为灵活的办公空间布局提供了先决条件。混凝土板随处可见，热量调节性能良好。一个整体的钢桁架为建筑前端的自由悬臂提供支撑。建筑的外立面绝大部分采用透明化处理，时刻确保室内光线充足。不透明的水平滑动式开窗为室内自然通风提供了有利条件。由一个金色钢丝网制成的机械化钢材元素完美地扮演了遮屏的角色。

与新建筑相对的原建筑一层空间也得到了改建。一个大型玻璃天棚为入口区进行了完美遮盖，其钢架结构与原建筑的内部结构遥相辉映。

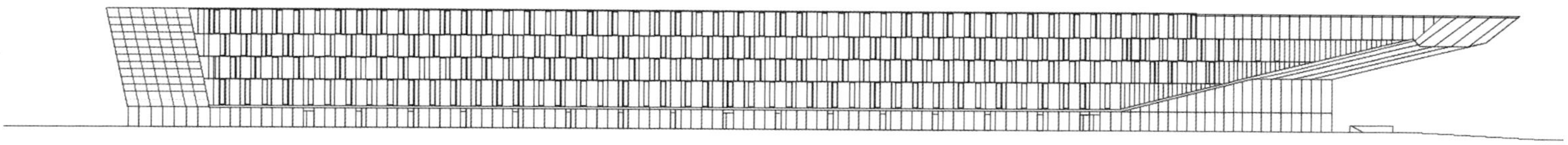

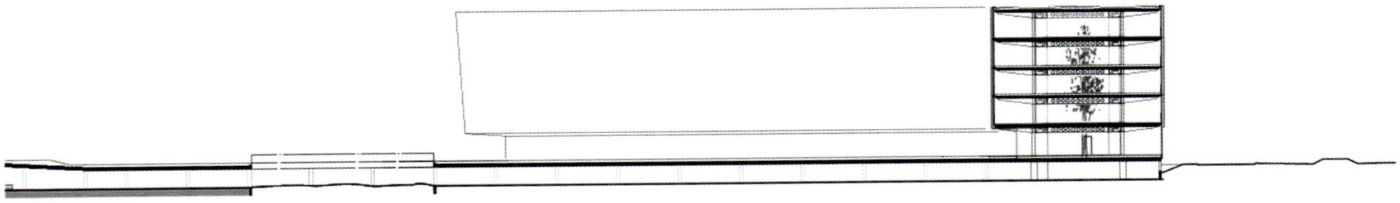

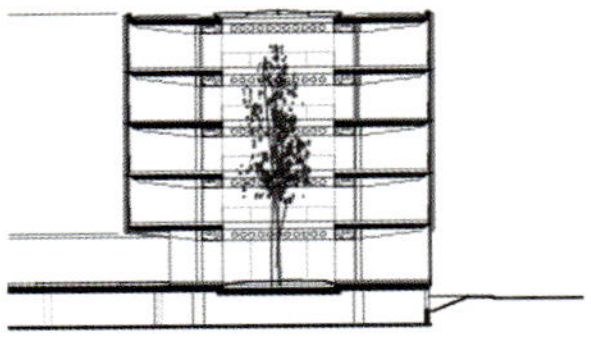

pine
voes

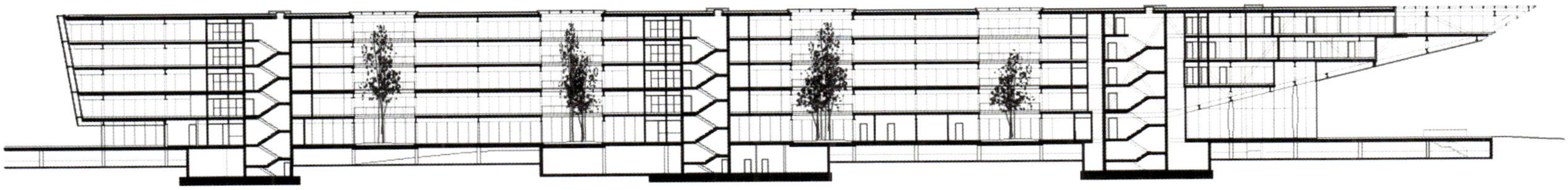

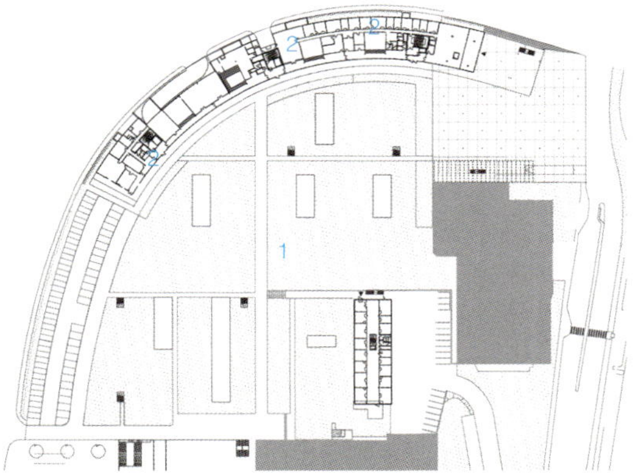

1. Parking 1. 停车场
2. Office 2. 办公室

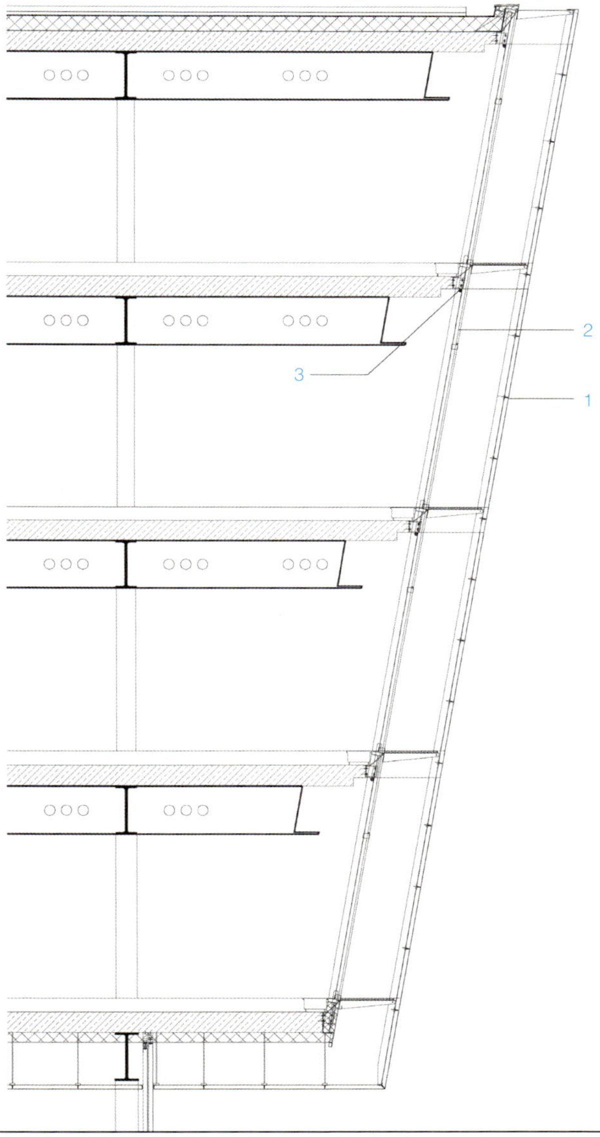

1. Sunshade fixed element of expanded metal
2. Stoery-high fixed glazing, glass type and dimension according to regulation and static
3. Glare protection

1. 金属网板遮阳固定结构
2. 高层固定玻璃，玻璃类型和尺寸根据规定和静力学而定
3. 遮光系统

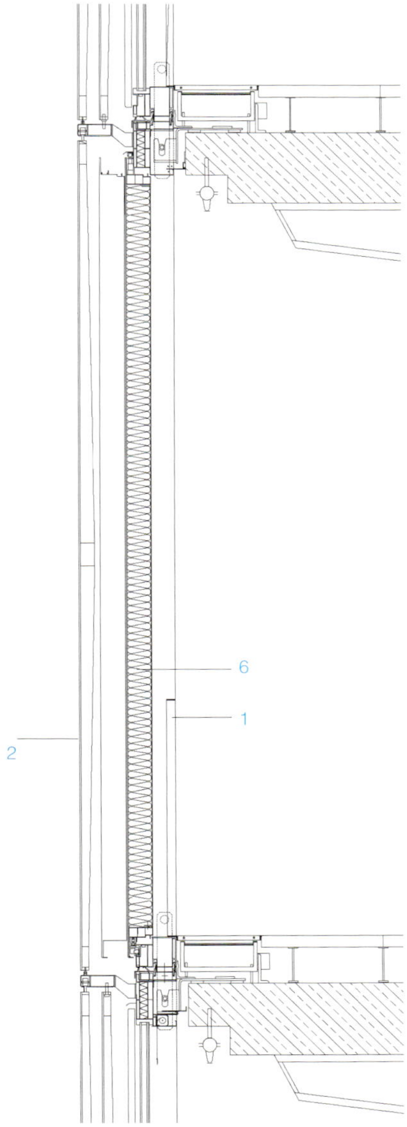

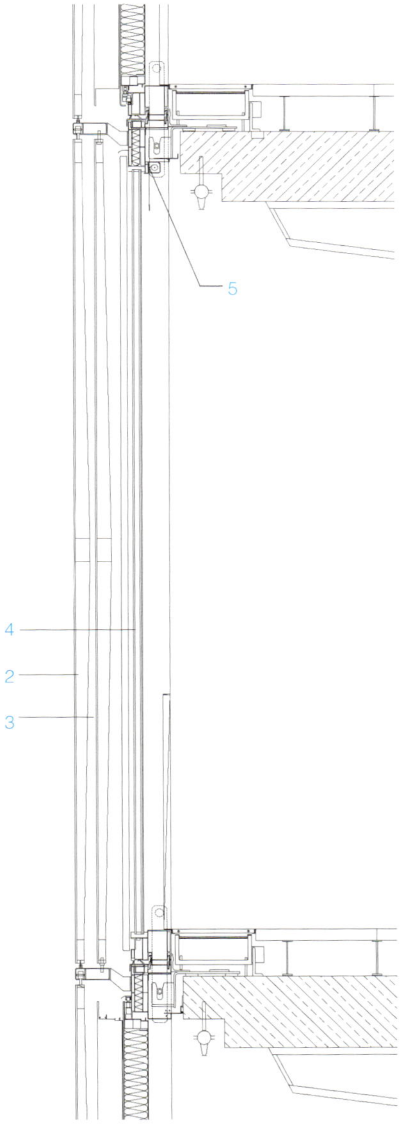

1. Flat steel safety device
   powder coated ral 7036
2. Sunshade sliding element of expanded metal on
   flat steel construction
3. Sun shade fixed element
   powder coated metallic yellow
4. Storey-height fixed glazing
5. Glare protection device
6. Parallel sliding window as sandwich panel
   powder coated ral 7036

1. 扁钢安全保护
   ral 7036粉末涂层
2. 扁钢结构上的金属网板遮阳滑动结构
3. 遮阳固定结构
   金属黄色粉末涂层
4. 层高固定玻璃
5. 遮光装置
6. 平行滑动窗作为夹层
   ral 7036粉末涂层

Audenasa

西班牙Audenasa办公建筑

Location: Noain, Spain
Designer: Vaillo & Irigaray + Eguinoa
Photographer: José Manuel Cutillas
Completion date: 2009

项目地点：西班牙，诺艾恩
设计师：Vaíllo & Irigaray + Eguinoa设计师事务所
摄影师：乔斯·曼努埃尔·科特拉斯
竣工时间：2009年

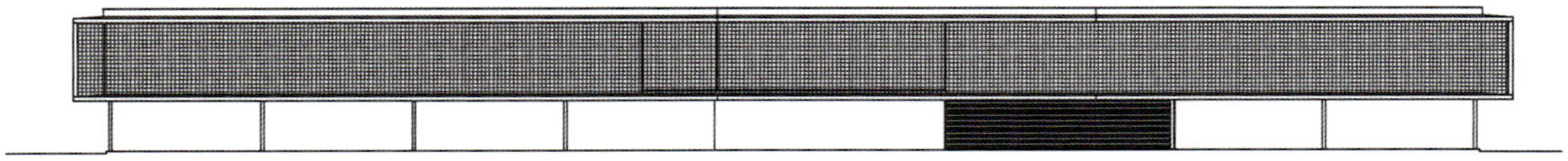

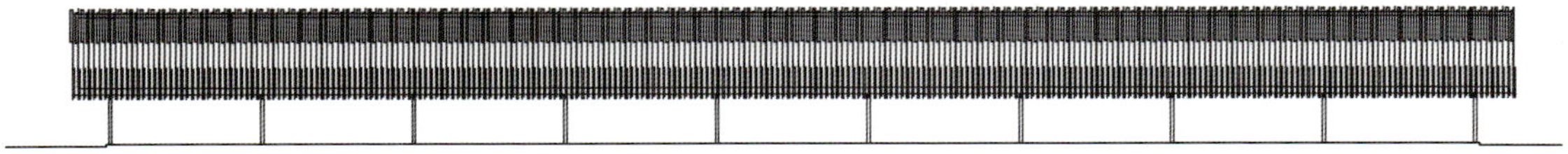

The planned location corresponds to a gentle slope facing south, located in the maintenance area called Noain AUDENASA. The programme consists of firms whose zoning responds to the various areas of the company, located on high ground and leaving the ground floor for access and covered parking. The organisational structure is configured using a central distribution hub: work areas are located south homogeneously oriented spaces and other "servants" are organised to the north.
The building offers an image derived morphological genesis of his own: a tablet suspended, almost floating on the gentle green slope, slightly twisted, repeating the same gestures as the topography-and offers a gesture of successive concave ribs against the sun. In a flat landscape, almost one-dimensional, as is the highway, immeasurably longitudinal, the building from where it controls and directs the company, contorts, and stands as lookout (also longitudinal), as a new "lookout" observer... Two slabs of concrete lattice steel tape the cor-ten blocks south and north reused tire. The picture of the complex aims to establish close ties to the movement and infrastructure relating to transport, and perhaps away from the usual urban readings in similar programmes.
This would give a line of work linked to making passive measures in environmental and bioclimatic consideration: in this case the orientation, respect for the topography and the inclusion of the building following the laws of the place are essential in working patterns:
South-facing slope: to protect from prevailing winds by the hill itself.
Opening to the south: maximum solar gain in winter and summer direct sun protection: lattice
Insulation: multilayer insulation façade of the north.
Adaptation geometric topography: the building twists to fit the contours of opting for a solution of minimum intervention and minimum landscape construction resources.
This result suggests a building parallel to the slope, longitudinal, glass and protected by a lattice on its southern side, closed tight and very isolated in the face of the north.

该项目位于诺艾恩Audenasa维修区一个坐北朝南的缓坡之上。建筑的上方按照公司的不同区域进行分区设计，下方的一层空间主要作为入口和停车处之用。项目的结构围绕一个中心分布区展开布局：工作区选择南向设计，而其他“服务区”则是面北而居。
建筑从形态学的角度来看，其形象是：悬垂的板块体量，漂浮在绿色的斜坡上，并且轻微弯曲，向阳面设有连续的凹面肋骨一般的钢架。在一个如同公路一样的一维平面景观之中，建筑向水平两端进行无限延伸，在为公司提供办公和管理空间的同时，如同一个哨所，时刻观察着，思考着……建筑的屋顶和地面分别是两块混凝土板；南面以科尔坦耐大气腐蚀高强度钢为主要原料；而北面则采用回收轮胎打造而成。建筑的外观旨在加强运动与交通设施之间的联系，并刻意地与其他类似的城市建筑项目进行鲜明地区分。
建筑的结构设计巧妙地将被动的环境措施与生物气候因素结合在一起：该方位设计有效地遵循了地形学原理，内部空间的布局同样根据工作模式进行设置。
南向的斜坡：抵御来自山中的盛行风。
南向的栅格：增加冬季太阳能的获取，避免夏季强光的照射。
绝缘层：北部设有多层绝缘立面。
几何形状外观：蜿蜒迂回的建筑形态能够将对景观的干预程度降到最小，尽量降低对景观建设资源的消耗。
因此，设计的方案促成了一个平行于斜坡、水平方向的建筑的诞生，其南端设置了玻璃和保护格栅，北部立面采用封闭独立式设计。

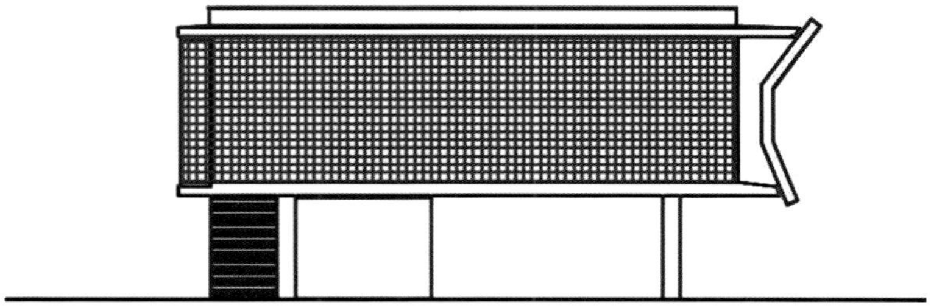

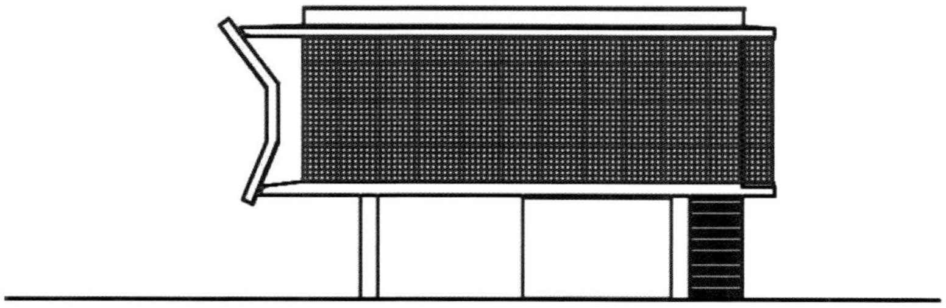

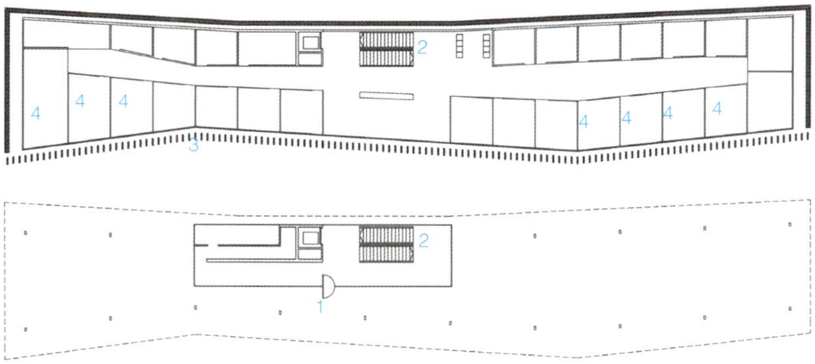

1. Entrance
2. Stairs
3. Corridor
4. Office

1. 入口
2. 楼梯
3. 走廊
4. 办公间

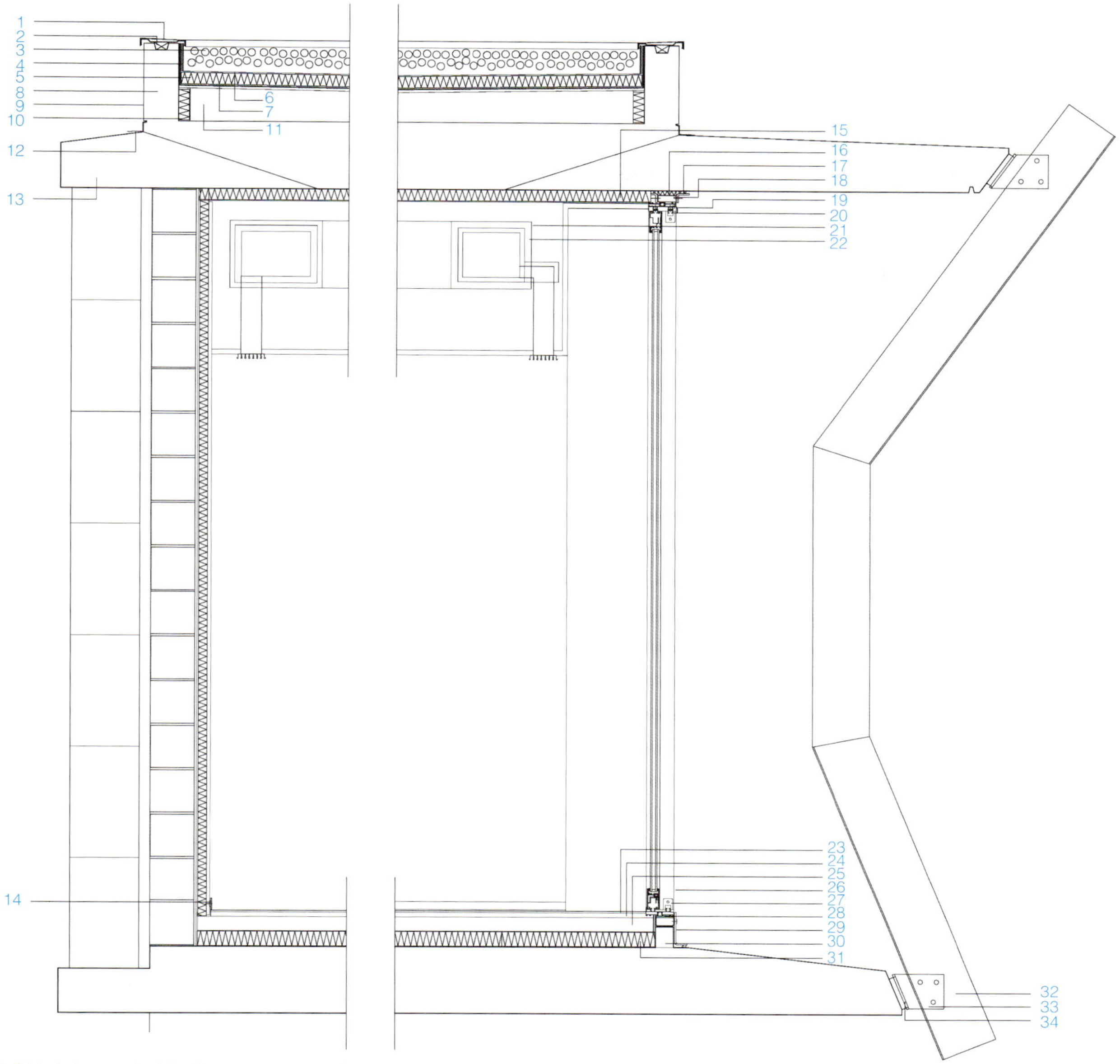

1. Folded plate top aluminium lacquer e=2mm type fastening clip with every 60cm
2. Wood stubs
3. Serious protection and aeration boulder D=20/30 mm
4. Plate protections
5. High density insulation dovetail roof mate e=70 mm
6. Geotextile anti-punching lamina
7. Two plates with modified bitumen SBS rubberised polymers of 2.4 Kg/m^2 and 4 Kg/m^2
8. Parapet concrete walls
9. Sikagard colourless priming type 710
10. Perimeter walls insulation roof mate separator with concrete slope
11. Concrete slope
12. Board of concrete sealing plate aluminium + bib
13. Clear primer type SIKAGUARD 710
14. Plinth crown lacquer solid aluminium e=8 mm bolted through wall of 5mm Teflon separator
15. Isolation of high density planned polyurethane e= 70mm, fire behaviour m1, painted black colour
16. Foam seal polyurethane
17. Bi-fold aluminium panel e= 6 mm
18. Galvanised steel frame 80-30-1.5
19. Sliding aluminium joinery
20. Double glazed with air spring 3+3/10/4+4 low emissivity
21. Rockwool insulation with aluminium interior vapour barrier, sealed e=50mm
22. Plaster cardboard interior partition on galvanised steel profiles e=45+15mm
23. Flooring: carpet heavy duty to traffic, m1, interface type
24. Base levelling mortar
25. Concrete sill e=50/70mm
26. Sliding aluminium joinery
27. Double glazed with air spring 3+3/10/4+4 low emissivity
28. Galvanised adjustable steel frame 80-40-3 (placement tolerance 1mm)
29. Bi-fold aluminium panel e= 6 mm
30. Concrete filled insulation projected non-retractable
31. Isolation of high density planned polyurethane e= 70mm
32. Lama folded plate formed by welded steel COR-TEN h=250mm e=3/4mm
33. Steel plate bracket COR-TEN blades e=8mm
34. Steel anchor plate COR-TEN e=10mm

1. 上方涂以铝漆的折叠板　E=2毫米，每60厘米处设置扣件
2. 木桩
3. 危急防护与通风卵石墙　直径为20或30毫米
4. 护墙板
5. 高密度绝缘屋面衬板e=70毫米
6. 土工布反冲孔板
7. 两个SBS改良型沥青防水卷材，型号分别为2.4千克/平方米，4千克/平方米
8. 混凝土护墙
9. 无色底漆类型SIKAGUARD710
10. 屋顶绝缘围墙与混凝土护坡分离开来
11. 混凝土斜坡
12. 混凝土密封木板＋铝材
13. 底漆采用SIKAGUARD 710材料
14. 基座上方涂漆的实心铝材e=8毫米，采用螺栓与5毫米厚的聚四氟乙烯隔板相衔接
15. 独立的高密度链接菱形聚氨酯板e=70毫米，耐火性能，黑色
16. 泡棉聚氨酯密封板
17. 双重折叠铝板e=6毫米
18. 镀锌钢架80-30-1.5
19. 滑动铝合金窗
20. 带气垫的双层低辐射玻璃　规格为3+3/10/4＋4
21. 内置铝材防潮层的岩棉绝缘层，密封e=50毫米
22. 镀锌钢型材上的内部石膏板间隔e=45+15毫米
23. 地板：重负载底板，接口类型
24. 水泥基自流平砂浆
25. 混凝土窗台e=50/70毫米
26. 滑动铝合金窗
27. 带气垫的双层低辐射玻璃　规格为3+3/10/4+4
28. 镀锌可调钢架80-40-3（公差1毫米）
29. 双重折叠板材e=6毫米
30. 混凝土填塞绝缘非伸缩板
31. 独立的高密度链接菱形聚氨酯板e=70毫米
32. 由高为250毫米，e为3/4毫米的焊接科尔坦耐大气腐蚀高强度钢制成的金属薄条折板
33. 科尔坦耐大气腐蚀高强度钢钢板支架e=8毫米
34. 科尔坦耐大气腐蚀高强度钢锚件e=10毫米

No. 5 Merchant Square

帕丁顿商业广场5号

Location: London, UK
Designer: Mossessian & Partners
Photographer: Nick Merrick
Completion date: 2010

项目地点：英国，伦敦
设计师：Mossessian & Partners建筑师事务所
摄影师：尼克·梅瑞克
竣工时间：2010年

No. 5 Merchant Square designed by Mossessian & Partners for European Land, is a 25,764 square metres (NIA) office building on a 3,855 square metres site within the Merchant Square development at Paddington, London.

Mossessian & Partners has deployed an architecture which is slim, dynamic and light, and creates a visually striking landmark building, appropriate to its context. It has sculpted the mass of the office block into three distinctive triangular shapes, each responding and mediating a relationship to the immediate context in all directions. To the West, they reflect the form of the existing buildings "Waterside" and "The Point"; to the South they provide a dynamic aspect to the canal; to the East they address the new public plaza; and to the North-West they secure a pedestrian circulation route towards the North Wharf Road and across it to a potential future development site.

The roof of the two "outside" triangles is cut back to create sloping south- facing roof planes. The central Triangle, coloured burgundy red, dramatically raises four storeys above its companions to punctuate the overall form and anchor the building in the general composition of the Basin. It also affords occupiers in the upper floors excellent views over the canal. No. 5 Merchant Square is further animated through the use of inward-sloping façades on the long edges of each triangle.

Mossessian & Partners has prioritised lightness and energy performance in designing the building's envelope. Each triangular element is cladded in a fully unitised aluminium framed curtain wall system, which varies to create a textured exterior appearance avoiding the possibility of a monolithic façade, or the general perception of a mute and dark glass office building.

The exterior of triangle 1 comprises of alternating rectangular units of clear, laminated and opaque glass, arranged at varying depths. The envelope of triangle 2 is made up of ceramic frit coated glass panels and includes operable louvres. A coloured dot pattern is incorporated, red on the exterior and black on the interior, to counteract the light and provide comfortable views out, while rendering a red expression from the outside. The design also allows for a pattern to be incorporated into the North façade of triangle 2. The theme of the glass is resumed around triangle 3 with the addition of precast white concrete expressing the slab edge. The roof is composed of alternating opaque panels of ceramic frit coated glass, skylights and includes provision for solar panels.

由Mossessian & Partners建筑师事务所为欧洲地产开发商设计的帕丁顿商业广场5号，是伦敦帕丁顿占地3,855平方米的商业广场中的一个办公建筑，实际用地面积达25,764平方米。

这个修长、光鲜、充满动感的地标型办公建筑为其所在区域增添了一道亮丽的风景。这个庞大的建筑由三个独特的三角形体量构成，各自与其所面向的周围景观完美地融合在一起。西向的体量与已有的“水乡”和“尖峰”大厦遥相辉映；南向的体量则与运河形成动态的和谐之美；而东向的体量与新建的公共广场相得益彰；最后，体量在西北方向时刻向来自于北仓道方向的行人致意，北仓道的对面是一个潜在的未来发展区。

两个“外侧”三角形的屋顶被巧妙地消减之后，形成了一个朝南的坡面。中央的三角体量，采用酒红色色调，比两侧的建筑高出四层空间，强调出整体形态的同时，并突出建筑在帕丁顿运河中的地位。此外，该部分体量还能够为居住在上层空间的住户提供更完美的运河景观。每个三角形体量狭长边缘地带的内向坡面外观令帕丁顿商业广场5号焕发出勃勃生机。

Mossessian & Partners建筑师事务所在该建筑的表面设计中充分考虑了亮度和节能的因素。每一个三角形元素均被覆盖在一个整体的铝材框架幕墙系统之中，该系统富于变化，打造了一个充满质感的建筑外观，避免了建筑立面的单一性，对传统的暗淡而厚重的玻璃办公建筑进行了大胆的突破。

第一个三角形体量的外部设置了交互式矩形单元，该单元由清晰的分层不透明玻璃按照不同的深度排列而成。第二个三角形体量由陶瓷烧结镀膜玻璃板材构成，其中还设置了可操作百叶窗，体量表面呈现出彩色光点图案，红色的外部和黑色调的内部将光线进行完美的中和，从外部投射到室内的红色调为室内营造了舒适美妙的景致。此外，该设计也与第二个三角形体量的北部立面完美融合。第三个三角形体量以玻璃为主题，边缘搭配白色混凝土预制板。其屋顶由交互式不透明陶瓷烧结镀膜玻璃板材、天窗和太阳能电池板构成。

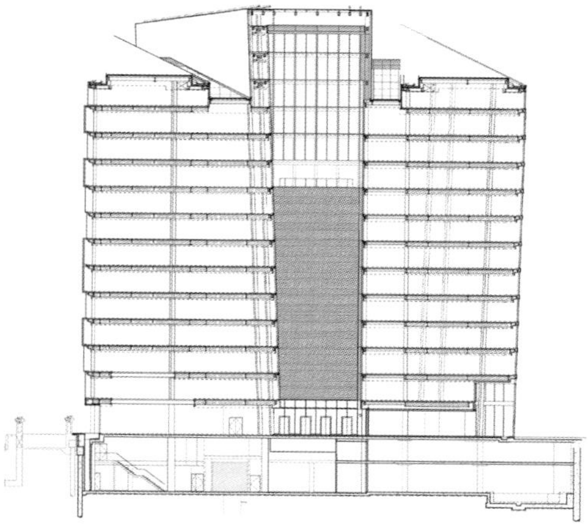

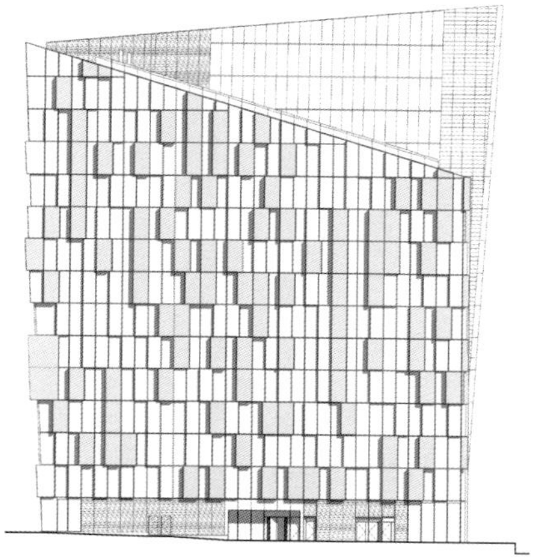

the waterline

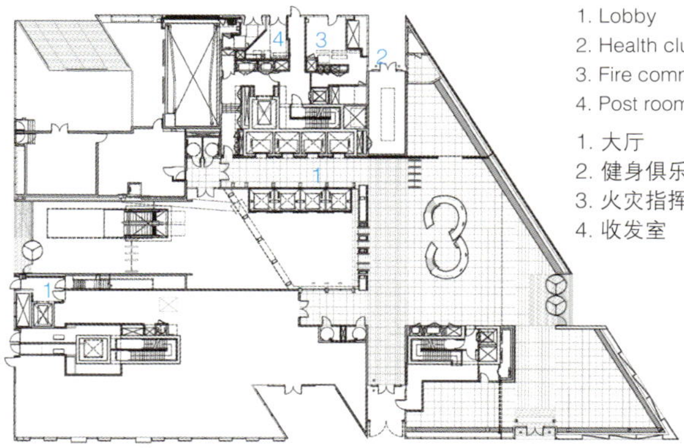

1. Lobby
2. Health club lobby
3. Fire command centre
4. Post room

1. 大厅
2. 健身俱乐部大厅
3. 火灾指挥中心
4. 收发室

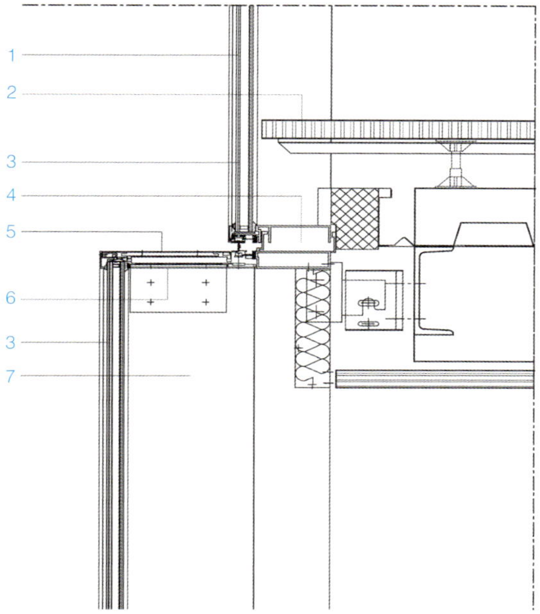

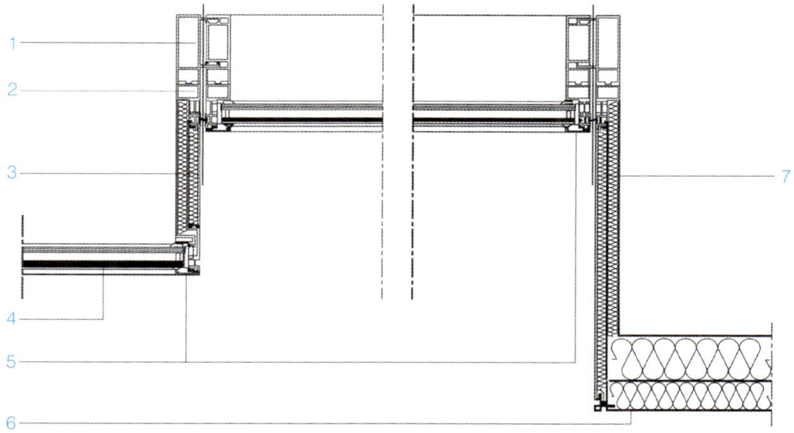

1. Low-E glass
2. Cantilevered floor system
3. Sand blasted glass screen
4. Unitised system - designed to accommodate extensions
5. Natural anodised finish
6. 250 mm extension mullion
7. Occupied interior space

1. 低辐射玻璃
2. 悬臂式楼面系统
3. 喷砂玻璃屏风
4. 组合系统－以适应结构的延伸
5. 天然阳极化抛光
6. 250毫米延长竖框
7. 使用中的内部空间

1. Unitised mullion - typical system wide clear anodised aluminium mullion
2. Unitised mullion extension attachment
3. Insulated / thermally broken extension mullion 250mm
4. Sandblasted glass screen
5. Identical capping conditions
6. Anodised aluminium plate
7. Insulated / thermally broken extension mullion 500mm

1. 整体竖框－标准的宽幅铝阳极氧化铝竖框
2. 整体竖框延伸连接物
3. 绝缘/热能分解延长竖框250毫米
4. 喷砂玻璃屏风
5. 完全相同的封顶条件
6. 阳极氧化铝板
7. 绝缘/热能分解延长竖框500毫米

City Hall of the Shanghai Nanhui District

上海南汇区市政厅

Location: Lingang New City, Shanghai, China
Designer: gmp – von Gerkan, Marg and Partners Architects
Photographer: Marcus Bredt, Berlin
Completion date: 2008

项目地点：中国，上海，临港新城
设计师：GMP建筑师事务所
摄影师：马库斯·布莱迪特，柏林
竣工时间：2008年

Size/Area: Gross floor area: 100,860 square metres
Partner: Nikolaus Goetze
Project leader: Jörn Ortmann
Design: Meinhard von Gerkan
Chinese partner practice: SIADR
Awarded date: 2005

尺寸/面积：总建筑面积：100,860平方米
合作伙伴：尼古劳斯·戈茨
项目负责人：乔恩·奥特曼
设计：冯·格康设计公司
中国设计搭档：SIADR设计师事务所
获奖时间：2005年

Nanhui District is now special in developing the Yangshan International Deep-water Port in the southeast and Pudong International Airport in the northeast. Yangshan Deep-water Port is a new port in Shanghai. It is connected the Lingang New City with the East China Sea Bridge. It can help Shanghai become the Top One harbour city in the world.

The Administrative Office Centre is embedded in the first green belt, which surrounds the city centre of Lingang New City. A river, which flows through, divides the complex into two separate areas. The prominent position of the plot by one of the main access roads and the important function for the district and the city demands a similarly exposed as well as strict architectural composition for the two individual buildings, which were grouped together to form a superordinate ensemble.

The high-rise at the west side of the site – facing the main entrance axis of the city – was designed as a landmark for Lingang New City. Two linear office buildings are forming an elongated block. At the eastern end they are joined to a round closure. In between the inner courtyard with its water basins and green trees two pedestrian bridges connect both parts of the complex, leading over the river.

The official access leads from the yard into the representative main lobby. Starting here, a cascade-like staircase leads up to the central zone, which is crowned by the conference hall with a capacity for up to 1,000 people. The public entrance for the inhabitants of the city and the Nanhui District is facing east – to the centre of Lingang New City and the lake.

并入上海市浦东新区之前的南汇区以重点发展东南侧的洋山国际深水港和东北部的浦东国际机场为产业发展的目标。通过东海跨海大桥与临港新城有机衔接在一起。这一目标将为推动上海成为世界第一大港口城市创造有利条件。

行政办公中心为环绕临港新城城市中心的首个绿化带打造了一道亮丽的风景。建筑被一条流经的河流分成两个独立的区域。独特的地理位置、行使的重要职能以及两个独立建筑严谨的结构共同打造了一个大气磅礴的建筑体量。

西侧高耸的建筑面向主入口的轴线设计，扮演了临港新城地标的角色。直线型办公楼群打造了一个纤长、高挑的办公空间。办公楼群在东端形成了一个圆形的封闭区。庭院中的水域、郁郁葱葱的树木、两处优雅的人行天桥将整体建筑的各个部分自然衔接在一起。

从庭院中沿办公大楼的入口可直接到达大堂。从此处开始，小瀑布式的楼梯将人们自然输送到中央区域，直至顶层的能够同时容纳1,000人的会议大厅。位于东侧的公共入口面向城市和南汇区居民开放，与临港新城城市中心和湖面遥遥相望。

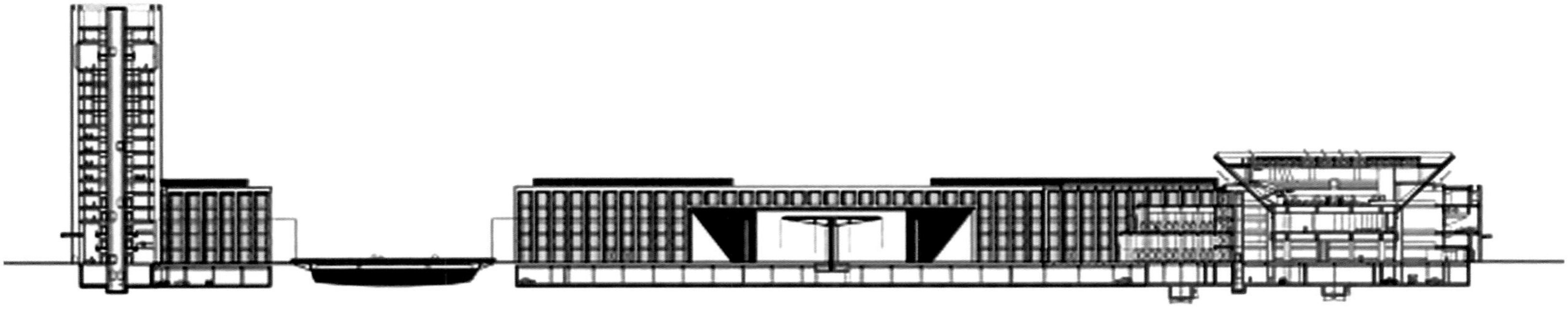

上海市南汇区人民政府
上海临港新城管理委员会

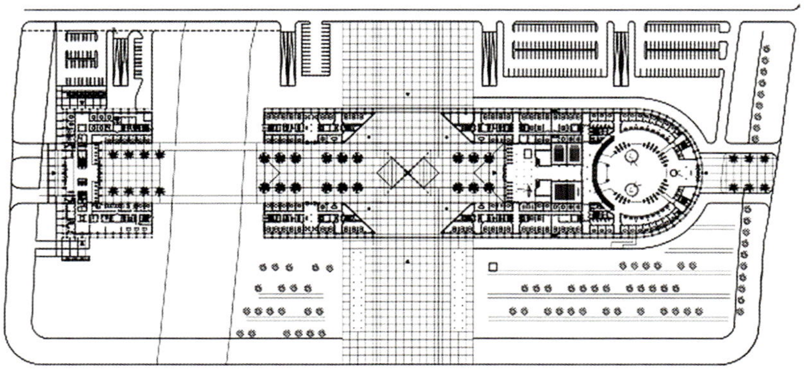

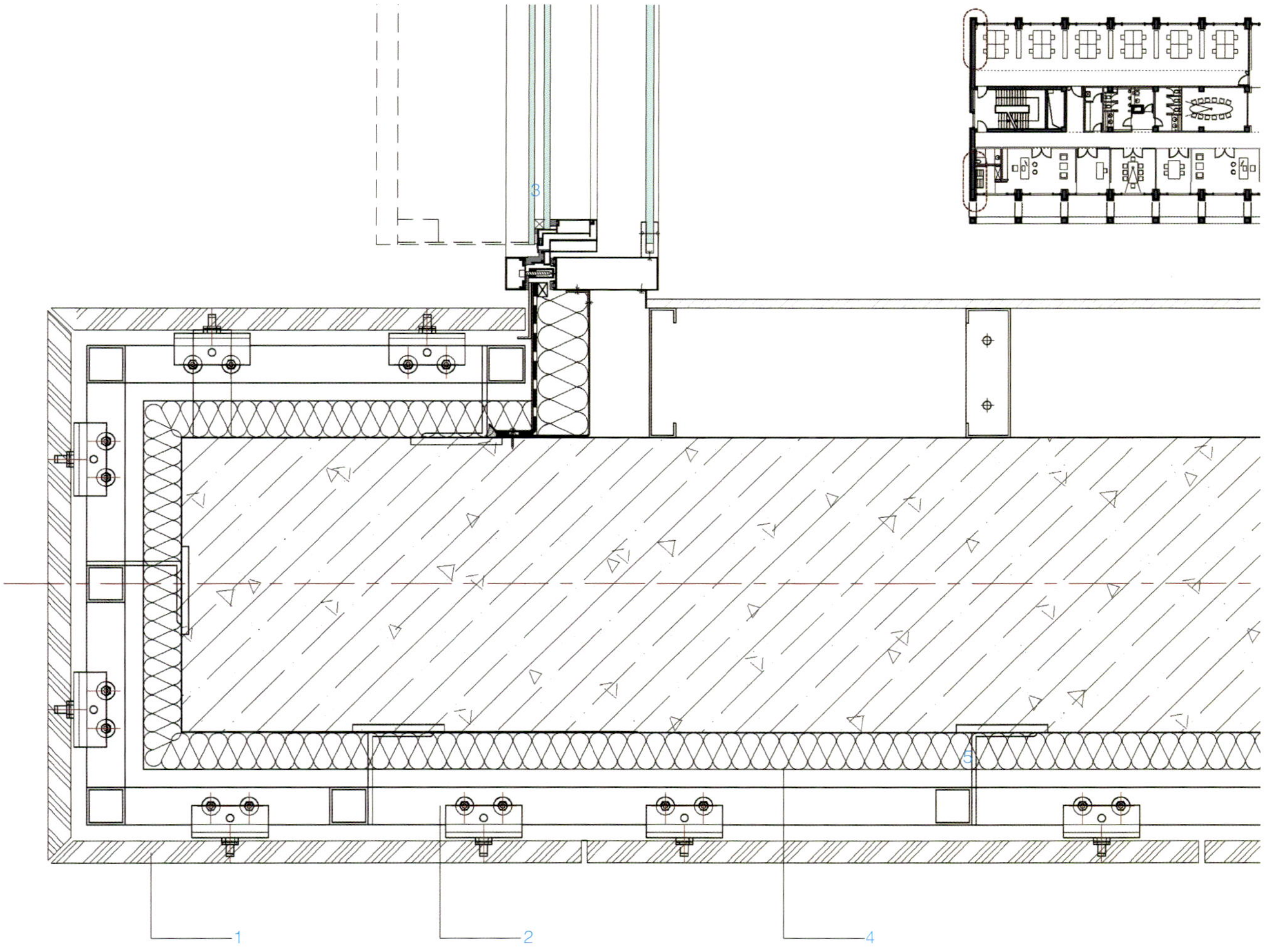

1. Natural stone, 30mm open gap
2. Aluminium frame with PVDF
3. Double layer Low-E glass
4. Insulation layer
5. Steel profile

1. 30毫米厚天然石材 开缝
2. 氟碳喷涂铝合金型材框
3. 双层中空低辐射镀膜玻璃
4. 保温层
5. 钢龙骨

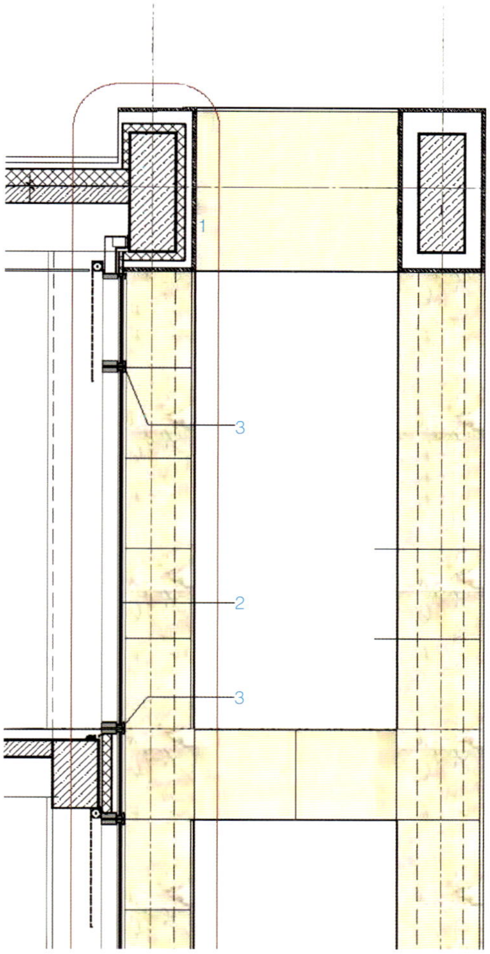

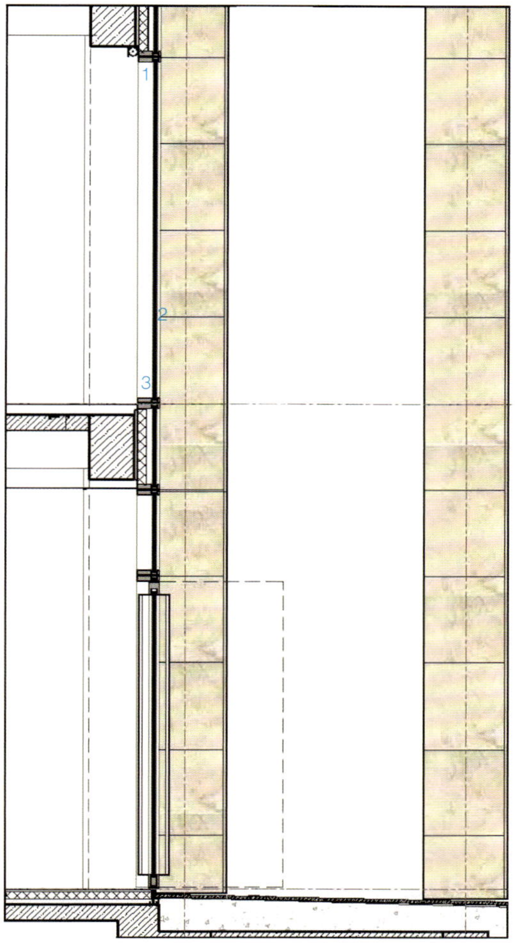

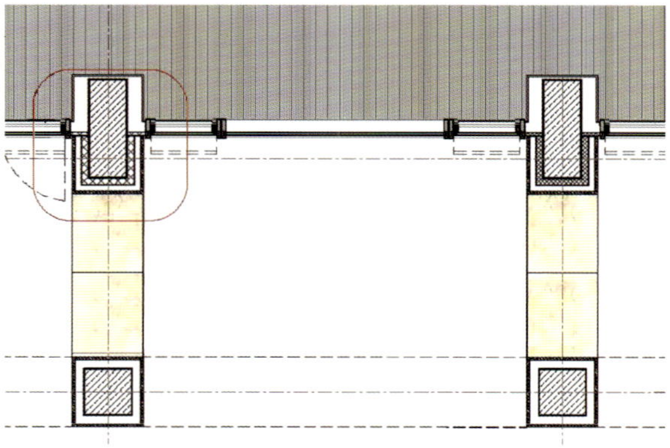

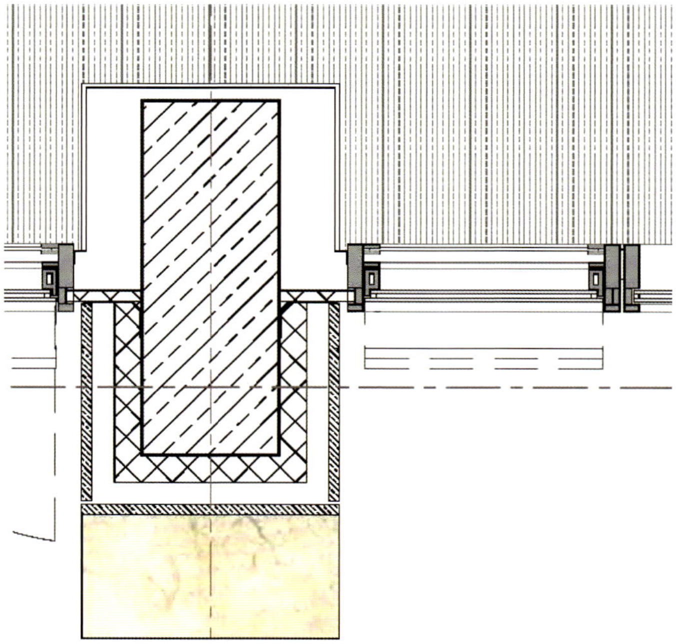

1. 30mm natural stone open gap
2. Double glass
3. Aluminium frame

1. 30毫米厚天然石材离缝干挂
2. 中空玻璃
3. 铝合金窗框

Aspis Bank Headquarters

Aspis银行总部

Location: Athens, Greece
Designers: Dimitris Papaioannou + Associates Architects, Planners, Engineers
Photographers: Nickolas Economou, George Papaioannou
Completion date: 2008

项目地点：希腊，雅典
设计师：Dimitris Papaioannou建筑师事务所等
摄影师：尼科拉斯·艾克诺，乔治·帕派安诺
竣工时间：2008年

Situated on a rather distinctive front, and on one of the busiest Avenues of Athens facing Eleftherios Venizelos Park, between the (listed) neoclassical Eginition Hospital and the residential blocks of the morphologically common urban fabric the scheme needed from the very beginning a design approach that would discreetly "link" its adjacent, though different, neighbors and retain the continuity of the frontage.

Keeping the above in mind two main volumes were formed, one at the height of the neighbouring neoclassical hospital and another one at the height of the adjacent buildings, giving significant emphasis on their verticality with the only conjunctive element their slim louvres.

However, the planning had to deal with all the technical problems, that were inherent in the particular site and which noteworthy often in the course of study required imperative solutions that rivaled intensely the firm objective for the substantial and not only morphological "integration" of the building. Most important was the direct adjacency with the Underground of Athens, as well as the existing circulatory congestion of the station and the Avenue.

The corner site, which has a clear rectangular form in plan, is fully exploited obtaining approximately a total of 5,000 square metres of useable space. The internal arrangement of spaces, followed by a thrift building programme, reflected the designers' clear and distinct awareness that the satisfaction of these programmatic requirements should clearly be readable externally. Through a simple and rigorous volume the Company that commissioned the project seeked to reflect a modern and dynamically developing appeal in the cityscape, immediately perceptible and clear.

Key themes in the design were the organisational symmetry of the floors, the simplicity of form and the utilisation of natural light. A fair void inside the building (atrium) lets in natural light via a series of skylights placed on the rooftop while "broadening" the internal circulation from top to bottom. Natural sand stone - Cyprus travertine - was used to clad the exterior and partially the interior of the building, and wood for the louvers.

Aspis银行总部位于雅典一条繁华的大街之上，与埃莱夫塞里奥・韦尼泽洛斯公园相对，介于新古典Eginition医院和住宅大厦之间，独特的外观与周围环境形成和谐的统一体，并凭借自身鲜明的设计特色而独树一帜。

建筑由两个体量构成，其中一个体量的高度与其附近的新古典主义医院持平，而另一个体量则力图与其毗邻的建筑相一致，建筑呈垂直形态布局，超薄百叶窗设计打造出连续的视觉效果。

然而，在设计过程中还涉及到诸多技术问题，这些问题是建筑独特的地理位置和研究过程中需要迫切解决的方案（诸如揭示公司运营的主旨或加强建筑形态上的统一等）中所固有的。最为重要的是建立与雅典地铁站以及原有车站和大街循环系统的直接衔接。

街角的位置，在平面布局上呈现出一个清晰的矩形形态，经过彻底开发后，其可用空间面积约为5,000平方米。内部的空间布局按照一个低调、严谨的建筑规划进行展开，有力彰显出设计师清晰、鲜明的布局理念，即力图体现空间布局的合理化、简洁化。通过一个简约、严谨的建筑完美彰显出该公司在城市景观中充满现代气息、动感之美的鲜明特色，清晰、婉约、亲切。

设计的关键主题在于楼层结构的对称性、形态的简约性以及自然采光的可行性。建筑内部的一个空地（中庭）通过一系列设于屋顶的天窗将自然光线成功引入室内，并"拓宽"室内上下空间的循环。天然砂石 – 塞浦路斯石灰华覆盖了建筑的外部表层以及室内部分区域和木制百叶窗。

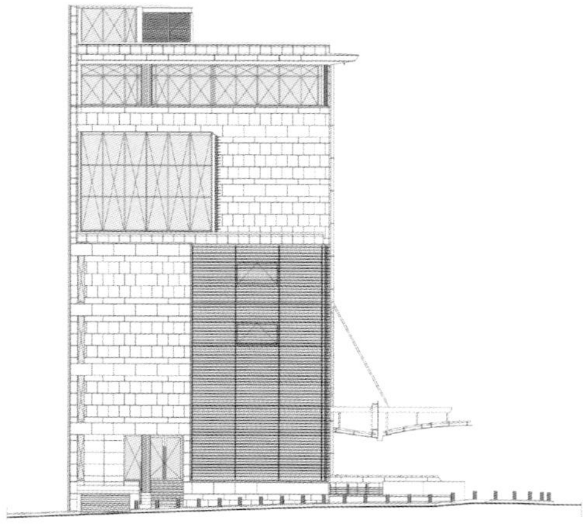

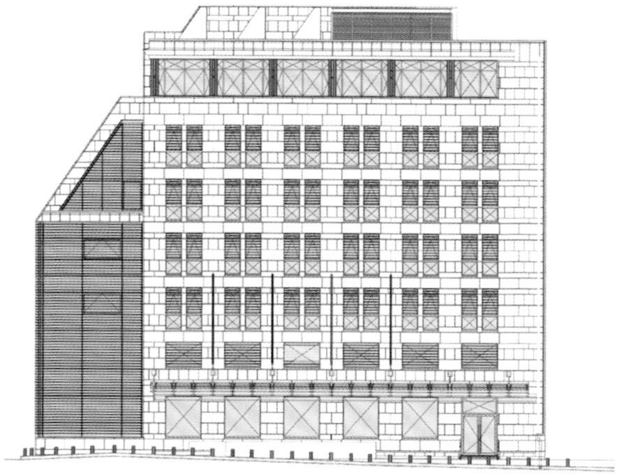

M

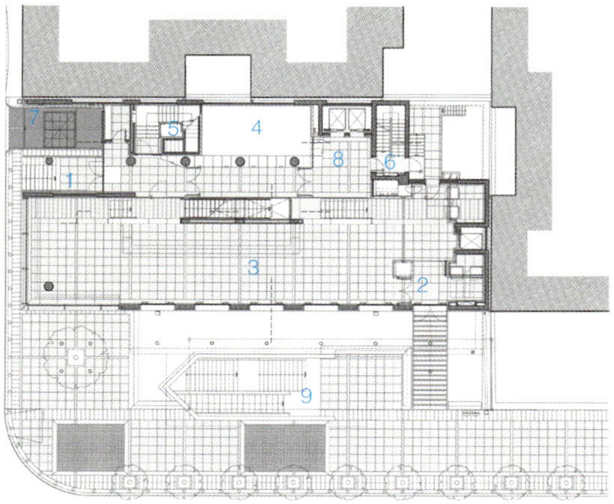

1. Entrance to offices
2. Entrance to bank
3. Retail banking
4. Atrium
5. Main stairwell
6. Fire escape staircase
7. Parking entrance
8. Elevators lobby
9. Metro-underground

1. 办公室入口
2. 银行入口
3. 小额银行业务
4. 中庭
5. 主楼梯间
6. 逃生楼梯
7. 停车场入口
8. 电梯大堂
9. 地铁

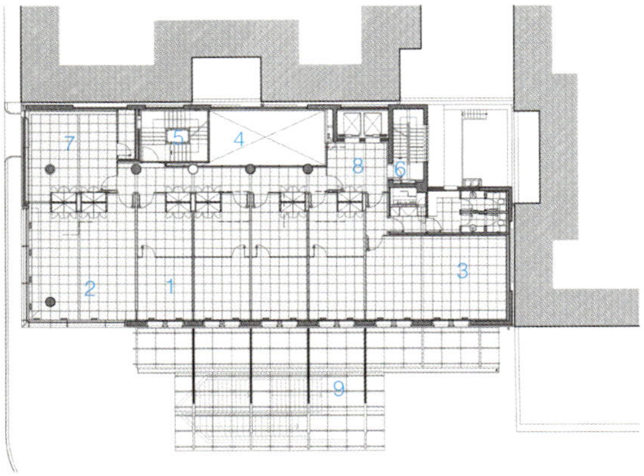

1. Typical offices
2. Offices
3. Offices
4. Atrium
5. Main stairwell
6. Fire escape staircase
7. Meeting room
8. Elevators lobby
9. Suspended glass canopy

1. 标准办公间
2. 办公间
3. 办公间
4. 中庭
5. 主楼梯间
6. 逃生楼梯
7. 会议室
8. 电梯大堂
9. 悬垂玻璃檐篷

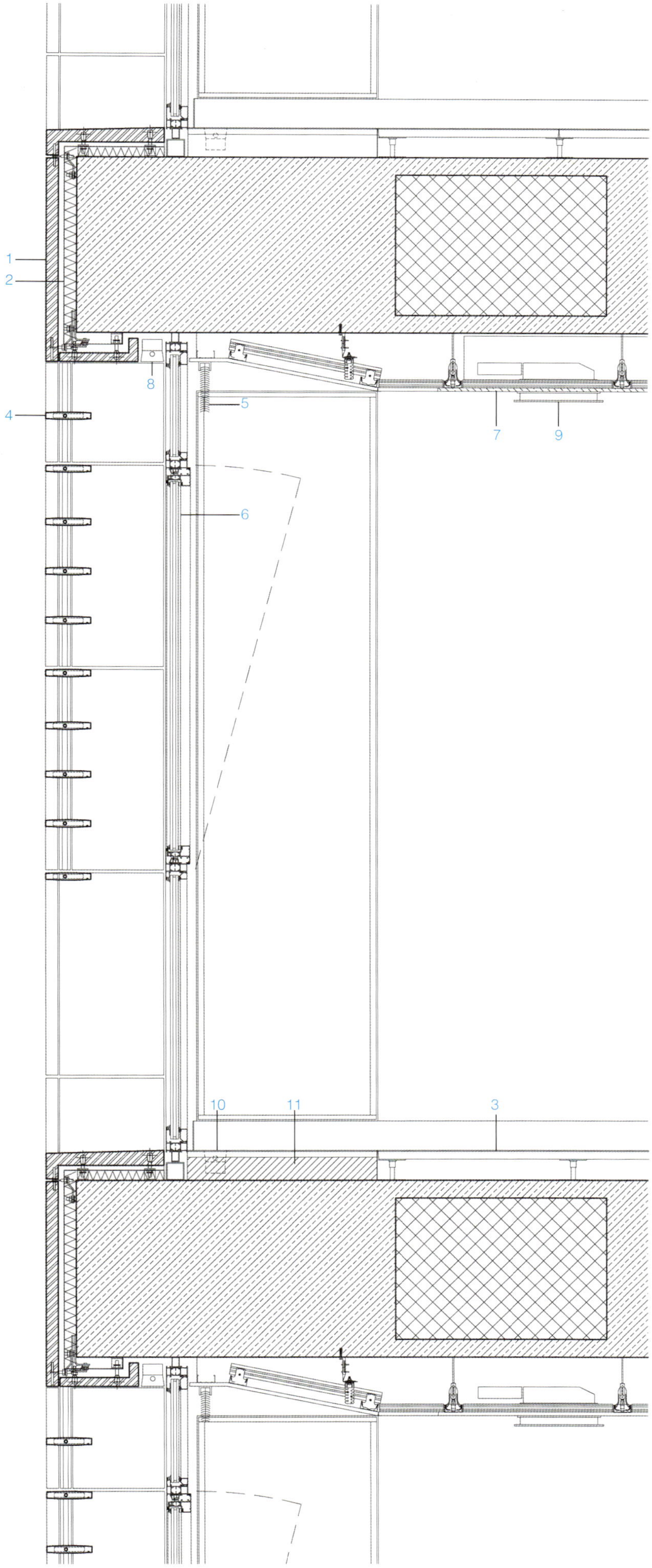

1. 40mm cyprus sandstone
2. 10mm void+50mm insulation
3. Office flooring:
   40mm double-layer floor slabs
   60mm void for electrical installation
   600mm reinforced concrete floor slab
4. 25mm wooden louvre
5. Aluminium louvre sunblind
6. Extruded aluminium casement and frame, thermally devided; glazing: 4mm+1.52mm PVB interlayer+4mm lam. Safety glass
7. 12.5mm gypsum plaster board on steel sections
8. External recessed WILA light fixture
9. Recessed WILA light fixture
10. In-ground WILA light fixture
11. 80mm cement screed

1. 40毫米塞浦路斯砂岩
2. 10毫米空隙＋50毫米绝缘层
3. 办公室地板：
   40毫米双层楼板
   60毫米电气安装空隙
   600毫米钢筋混凝土楼板
4. 25毫米木板散热孔
5. 铝材百叶窗
6. 挤压铝窗户和窗框，隔热；玻璃：4毫米＋1.52毫米聚乙烯醇缩丁醛夹层＋ 4 毫米安全玻璃
7. 型钢上的12.5毫米石膏灰泥板
8. 嵌入式灯具装置的外部
9. 嵌入式灯具装置
10. 地面灯具装置
11. 80毫米水泥砂浆底层

Monaco House

摩纳哥综合办公大楼

Location: Melbourne, Australia
Designers: Robert McBride, Debbie-Lyn Ryan
Completion date: 2007
Floor area: 400m^2

项目地点：澳大利亚，墨尔本
设计师：罗伯特·麦克布莱德，黛比·林恩·瑞安
竣工时间：2007年
占地面积：400平方米

This is a new four-storey building located in a largely pedestrian lane (Ridgway place) at the East end of Melbourne's CBD.
The brief was to provide a ground level entry and café, followed by two levels of office tailored for the Proprietors Investment and Philanthropic Organisation. The top level contains a small reception area primarily for official functions associated with the client's role as Honourary Consular of Monaco.
The client had a love of the design of cars, boats (particularly early the 20th) and finely crafted objects. He bemoaned the loss of shape in the contemporary world. It was in the area of shape, craft and material that the architect and client found the common ground.
To meet cost constraints building elements had to be structured around crane sizes, and crane sizes around turning circles. Much of the external is pragmatic and unremarkable. The designers wanted the building to be both abstract and awash with imagery. They looked at the plane trees, the gothic, surrealism, the heraldic, deco and the Prague cubists. They wanted the building to be above all else something that amplified its miniature urban grain and enriched the pedestrian experience of the city.

Within the office large apertures to the West are well shaded by deep balconies and the adjoining plane trees. The workspace has good natural light and cross ventilation. Exposed windows have electronically controlled external blinds. Outdoor balconies provide areas of release from the office desk. The "green roofscape" is a similar space but also adds additional insulation to the upper floor. Water is harvested and stored in the stairwell. Fixtures are selected for their low energy and water usage. There is a heavy emphasis on insulation with a combination of closed cell Styrofoam, air cell and bulk insulation reducing reliance on conditioned air.
From a technical point of view the complex geometry and use of such materials as Glass Reinforced Cement and its structural support demanded sophisticated use of 3D software and integration of those techniques with the off site production.

这个四层综合型办公建筑位于墨尔本中央商务区东区的一条步行大街上（里奇苇大街）。
建筑的设计理念是在一层空间打造入口大厅和咖啡馆，并在其上为资产投资和慈善机构提供两层的办公空间。建筑的顶层设有一个小型的接待中心，主要为摩纳哥名誉领事等客户提供接待服务。
建筑的用户对汽车、轮船（尤其是20世纪之前的）设计以及其他制作精良的工艺品情有独钟，并对当今世界形态的缺失而深感遗憾，而这个造型独特、设计精巧、材料精良的建筑使用户和建筑师达成了共识。
为尽量地节省开支，建筑的构建必须围绕起重机的尺寸以及起重器转弯半径进行构建。建筑的外部设计遵循实用的原则，视觉效果并不显著。打造抽象而形象化的建筑是该设计的宗旨。建筑巧妙地集超现实主义哥特艺术与布拉格立体派艺术风格于一体，作为城市宏观环境下的一个微观体现，并为行人提供完美的步行体验。

在办公空间的内部，西向的大型开口被狭长的阳台及其附近的梧桐树完美遮蔽。工作区采光和通风性能良好。外墙窗口设有电子自动控制百叶窗。室外露台为忙碌的办公环境提供优雅的放松空间。绿化屋顶与室外露台类似，相比之下，顶层增设的隔离空间则更为私密、幽静。楼梯井中可作为蓄水之用。装置的选用遵循节能、节水的原则。封闭式泡沫聚苯乙烯塑料与通风和体积绝热设计完美结合，有效地减少了对空调的依赖。
从技术角度来看，复杂的几何型外观和玻璃纤维强化水泥材料及其结构支撑物需要3D软件和其他技术的综合运用。

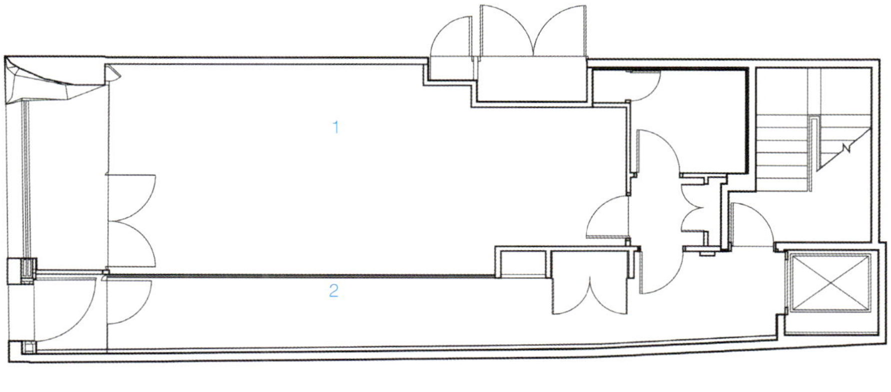

1. Café
2. Foyer

1. 咖啡馆
2. 门厅

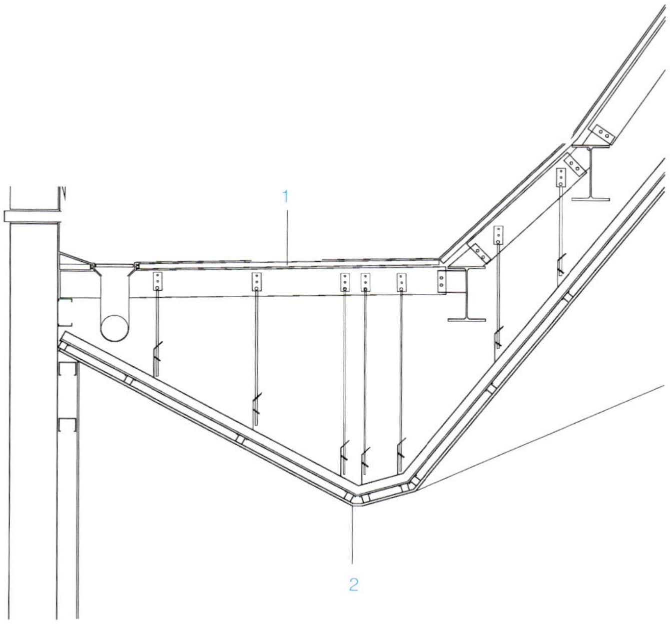

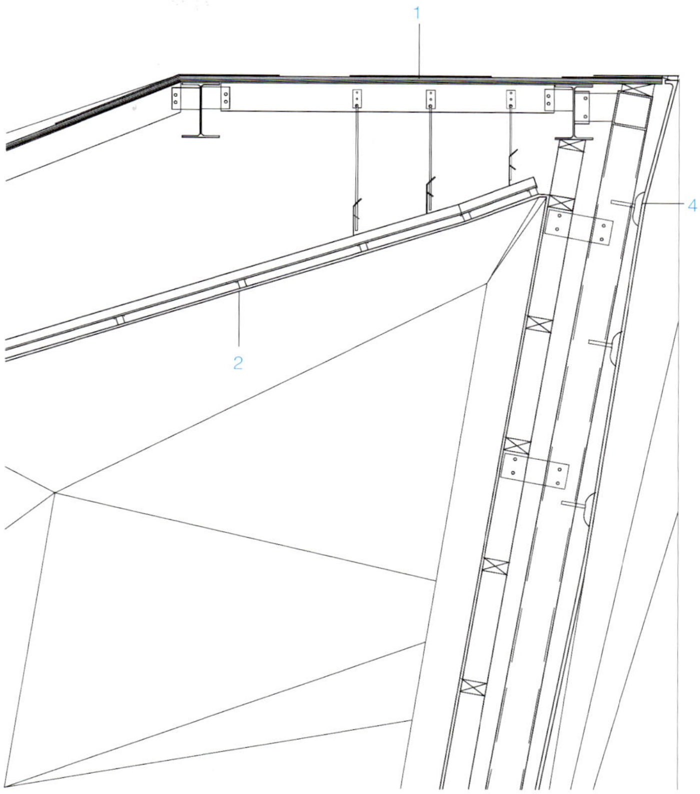

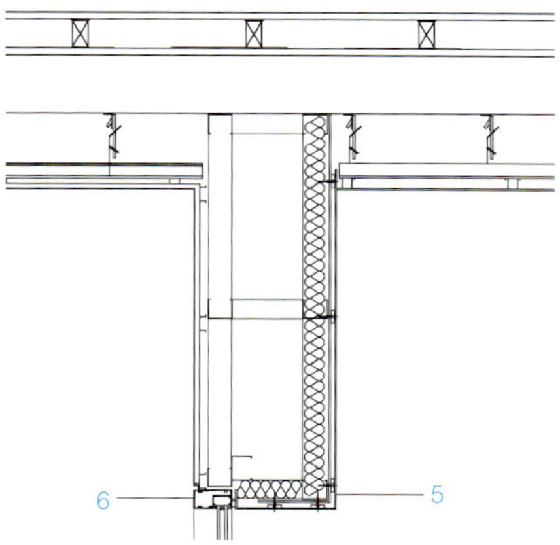

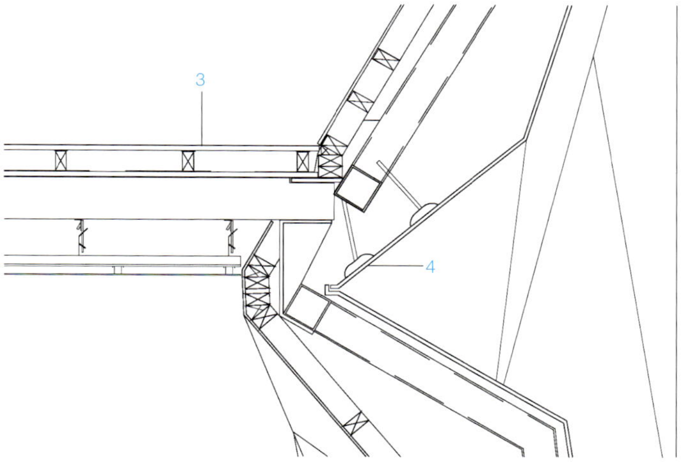

1. 1.5mm thickness ardex butynol membrane fixed to 19mm thickness marine plywood on steel roof battens
2. Painted and flushed 9mm "villaboard" (external) and plasterboard (internal) on suspended ceiling system
3. Turpentine hardwood decking with marine-grade wood oil finish; on kdhw timber battens on fortecon rubber packers, 1.5mm thickness ardex butynol membrane on concrete screed
4. Glass reinforced concrete façade cladding, 12mm thickness cast-in steel flex-anchors welded to steel support structure. shs perimeter, framing for grc wall
5. R2.5mm insulation batts behind fc sheet & aircell "retroshield" sarking, 3mm thickness ss plate countersunk screw-fixed to 6mm fc sheet on metal stud framing
6. Double glazed windows with aluminium frame flush with wall

1. 1.5毫米厚的亚地斯薄膜，固定于钢架屋顶吊杆上的19毫米厚的海洋胶合板之上
2. 吊顶系统中着色冲洗 9毫米的“水泥板”（外部）和石膏板（内部）
3. 松脂硬木由海洋龙脑香油树脂装饰而成；位于橡胶封隔器上方的木材板条之上；位于混凝土砂浆底层的1.5毫米厚的亚地斯薄膜
4. 12毫米厚的玻璃纤维增强混凝土外墙包层。现浇钢焊接的锚钢支撑结构。GRC墙体的自蔓延高温合成框架
5. 半径为2.5毫米的绝缘板条位于硅酸钙板和透气屋面衬板的后侧；3毫米厚的不锈钢板埋头螺丝钉固定在6毫米厚的金属立杆框架之上
6. 双层玻璃窗与和墙体平齐的铝框

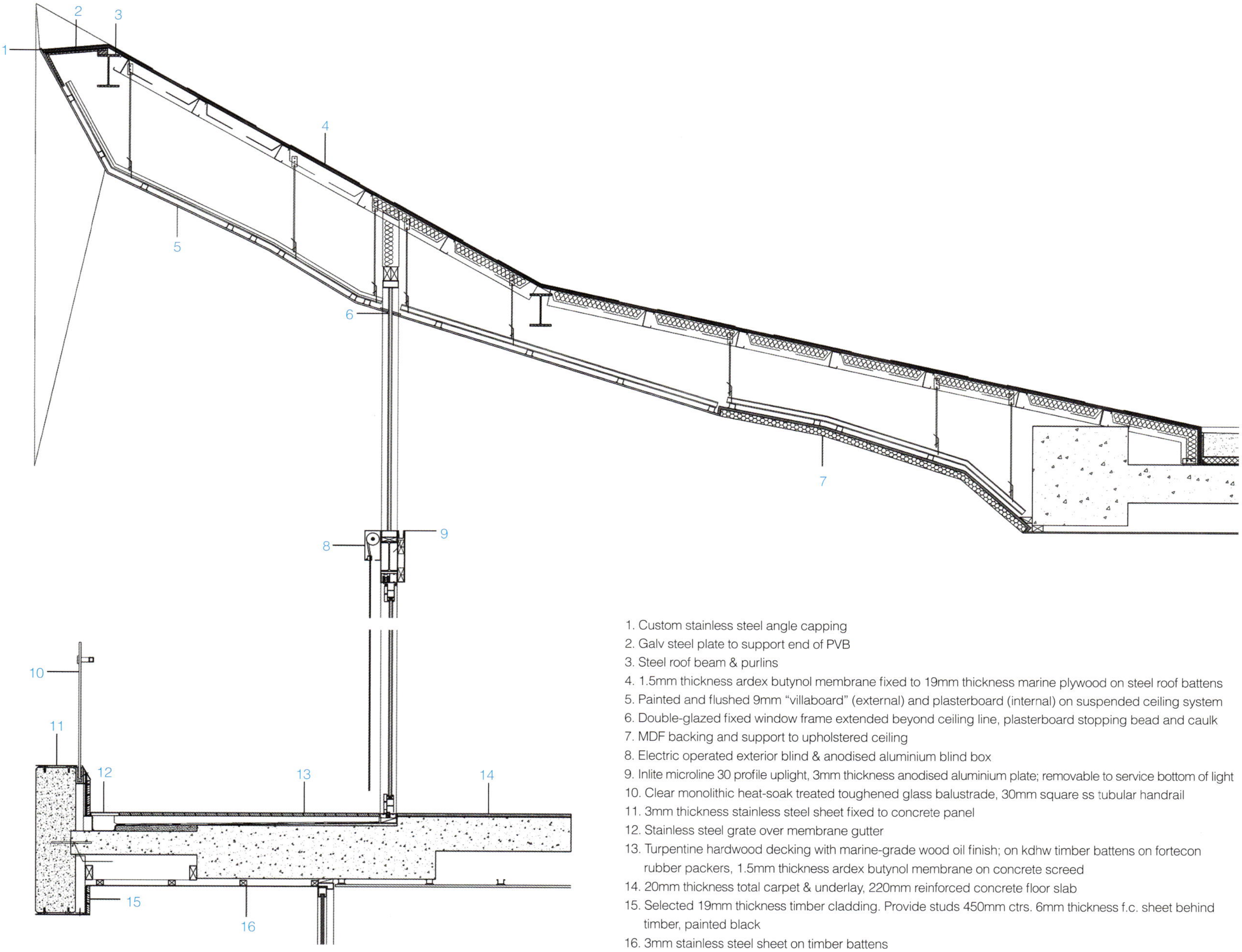

1. Custom stainless steel angle capping
2. Galv steel plate to support end of PVB
3. Steel roof beam & purlins
4. 1.5mm thickness ardex butynol membrane fixed to 19mm thickness marine plywood on steel roof battens
5. Painted and flushed 9mm "villaboard" (external) and plasterboard (internal) on suspended ceiling system
6. Double-glazed fixed window frame extended beyond ceiling line, plasterboard stopping bead and caulk
7. MDF backing and support to upholstered ceiling
8. Electric operated exterior blind & anodised aluminium blind box
9. Inlite microline 30 profile uplight, 3mm thickness anodised aluminium plate; removable to service bottom of light
10. Clear monolithic heat-soak treated toughened glass balustrade, 30mm square ss tubular handrail
11. 3mm thickness stainless steel sheet fixed to concrete panel
12. Stainless steel grate over membrane gutter
13. Turpentine hardwood decking with marine-grade wood oil finish; on kdhw timber battens on fortecon rubber packers, 1.5mm thickness ardex butynol membrane on concrete screed
14. 20mm thickness total carpet & underlay, 220mm reinforced concrete floor slab
15. Selected 19mm thickness timber cladding. Provide studs 450mm ctrs. 6mm thickness f.c. sheet behind timber, painted black
16. 3mm stainless steel sheet on timber battens

1. 特制的不锈钢角钢封顶
2. 镀锌钢板支撑PVB管尾
3. 钢屋盖梁和檩条
4. 1.5毫米厚的亚地斯薄膜，固定于钢架屋顶吊杆上的19毫米厚的海洋胶合板之上
5. 吊顶系统中着色冲洗 9毫米厚的“水泥板”（外部）和石膏板（内部）
6. 双层玻璃固定窗框延伸到天花线之外，石膏板窗止条与堵缝
7. 中密度纤维板衬垫与装饰天花板支撑结构
8. 电动操作室外百叶窗及阳极氧化铝百叶窗匣
9. 微斜石30侧面采光板，3毫米厚的铝阳极板；可移动，为下方空间提供光线
10. 整体单片热浸处理的钢化玻璃栏杆，30×30毫米不锈钢管扶手
11. 3毫米厚的不锈钢板固定在混凝土板上
12. 不锈钢格栅位于薄膜排水沟之上
13. 松脂硬木由海洋龙脑香油树脂装饰而成；位于橡胶封隔器上方的木材板条之上；位于混凝土砂浆底层的1.5毫米厚的亚地斯薄膜
14. 20毫米厚的地毯及底衬，220毫米钢筋混凝土楼板
15. 精心挑选的19毫米厚的木料覆层，双头螺栓450毫米；6毫米厚黑色硅酸钙板位于木板后方
16. 木材板条上的3毫米厚不锈钢薄板

Oporto Vodafone Building

波尔图沃达丰大楼

Location: Porto, Portugal
Designers: Barbosa & Guimarães (josé antónio barbosa, pedro lopes guimarães)
Photographer: Paulo Lima
Completion date: 2009

项目地点：葡萄牙，波尔图
设计师：Barbosa & Guimarães建筑事务所
摄影师：圣保罗·利马
竣工时间：2009年

Four years after the conclusion of Vodafone Lisbon headquarters, Vodafone decided to build a new building, at Porto, which allows concentrate in only one place for their workers.
The functional programme includes office areas, mega shop store, auditorium, cafeteria, training rooms, warehouses, technical areas and parking. The building is located in the corner of the Avenue Avenida da Boavista with the street Rua Correia de Sá, on a plot area with 1,970 square metres, facing the previous named streets at north and a garden at south, upper bound, where exist a few tree species that were preserved. The main volume of the building, presents fairly to Boavista Avenue, an altitude range between three and five floors above the threshold, according to high level of adjacent buildings.
This monolithic irregular volume, which looks to convey movement sensation, is bounded by walls and roofs with irregular and fragmented geometry, in accordance to the alignments defined by the dominant buildings on the east and west of the building. The building is vertically organised on eight floors, five above ground and three on the basement.
The ground floor, where lies the three accesses points from Boavista Avenue, is occupied by the lobby, mega store on two floors, auditorium and cafeteria. The auditorium and cafeteria are connected to the garden through the back courtyard and stairs. In the four next upper floors, are located the open space offices, two toilets and a pantry on each one of it. Natural lighting of the offices is carried out through continuous windows throughout the length of north and south elevations.
On the first lower floor, in addition to the extension of the store, are the training rooms, while the rest area occupied by warehouses and technical areas. The two following lower floors are destiny to car parking.
Structurally, the building is based on full concrete solution, with solid slabs supported on walls, cores, and some columns. The shell structure of white concrete self-compacted on the north and south elevations. The exterior walls were coated, internally, with plasterboards to form air boxes thermal insulates. The roof cover conclusion was made using white pre-fabricated slabs, on thermal insulation and waterproofing.

在移动通信运营商沃达丰公司里斯本总部竣工的四年后，该公司的另一座新建筑在波尔图完美落成，整个建筑专属为沃达丰公司，完整的办公环境为其员工带来了崭新的工作体验。
这一功能型建筑内设有办公区、时尚店铺、礼堂、自助餐厅、培训室、储藏室、技术区与停车场。建筑位于Boavista大道与Rua Correia de Sá大街的转角处，占地1,970平方米，北侧面向大街，南侧设有花园，其间种植的一些树木属于保护性树种。建筑的主体量面向Boavista大道，共八层，与周边的建筑高度相和谐。
这个庞大的不规则体量传达出动感之美。不规则的墙体和屋顶形态被大楼东西两侧的建筑界定出来，从而实现了视觉上的和谐、统一。呈垂直形态布局的大厦有8层，其中3层设于地下。
面向Boavista大道的地上一层设有三个入口，内设入口大厅、两层店面、礼堂以及自助餐厅。礼堂和自助餐厅经由后庭院和楼梯与花园完美衔接在一起。楼上四层空间中为公共办公区，每个办公间中设有两个卫生间和一个备餐间。贯穿南北立面、有序衔接的玻璃窗时刻确保室内环境通透、光线充足。
地下一层空间中除商店延伸部分以外为培训室，其余的空间则作为储藏室和技术区。地下二三层则为停车场。
从结构上看，建筑全部采用混凝土结构，墙体、核心以及一些梁柱由实心板进行支撑。南、北立面的外部结构采用白色自密实混凝土材料。涂层外墙内部设置石膏板，能够进行完美隔热。屋顶采用白色预制板材料，具有理想的隔热和防水的功能。

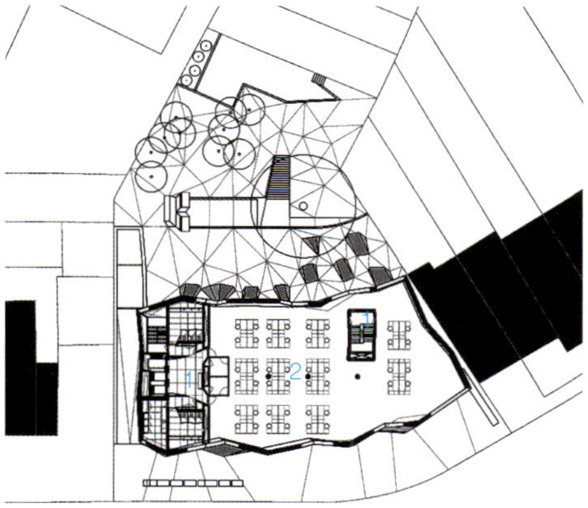

1. Distribution
2. Open space

1. 配送区
2. 开放空间

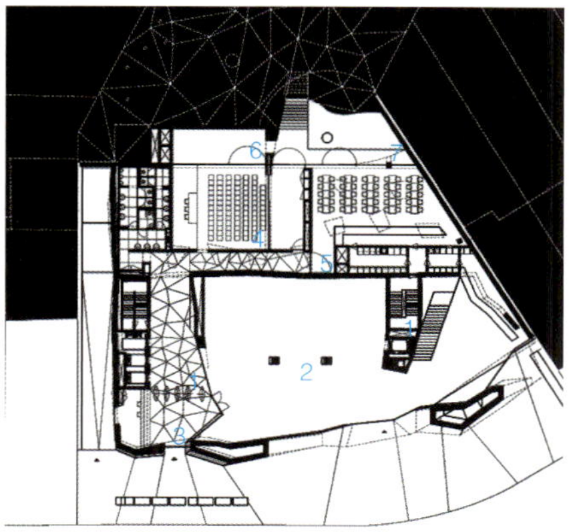

1. Distribution
2. Megastore
3. Office access
4. Auditorium
5. Cafeteria
6. Courtyard
7. Garden

1. 配送区
2. 购物广场
3. 办公区入口
4. 礼堂
5. 自助餐厅
6. 天井
7. 花园

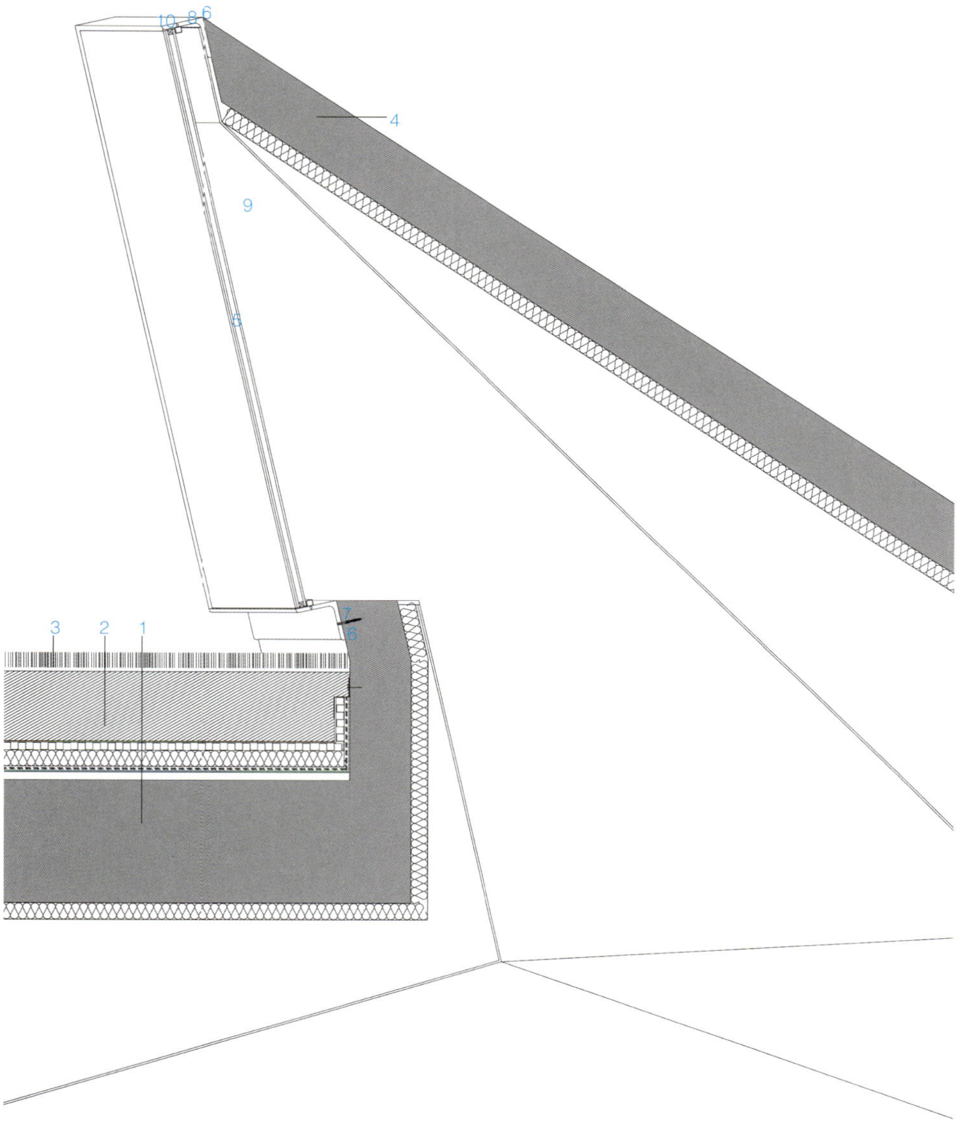

1. Concrete slab
2. Humus
3. Grass
4. White concrete
5. Glass 4+4+12+6mm
6. Steel bracket to paint
7. Metal bushing
8. Tubular steel 15cmx15cm to paint
9. Plaster board
10. Structural silicone

1. 混凝土板
2. 腐殖质层
3. 草坪
4. 白色混凝土
5. 玻璃4毫米+4毫米+12毫米+6毫米
6. 喷涂钢支架
7. 金属衬套
8. 喷涂钢管15厘米x15厘米
9. 石膏板
10. 硅结构

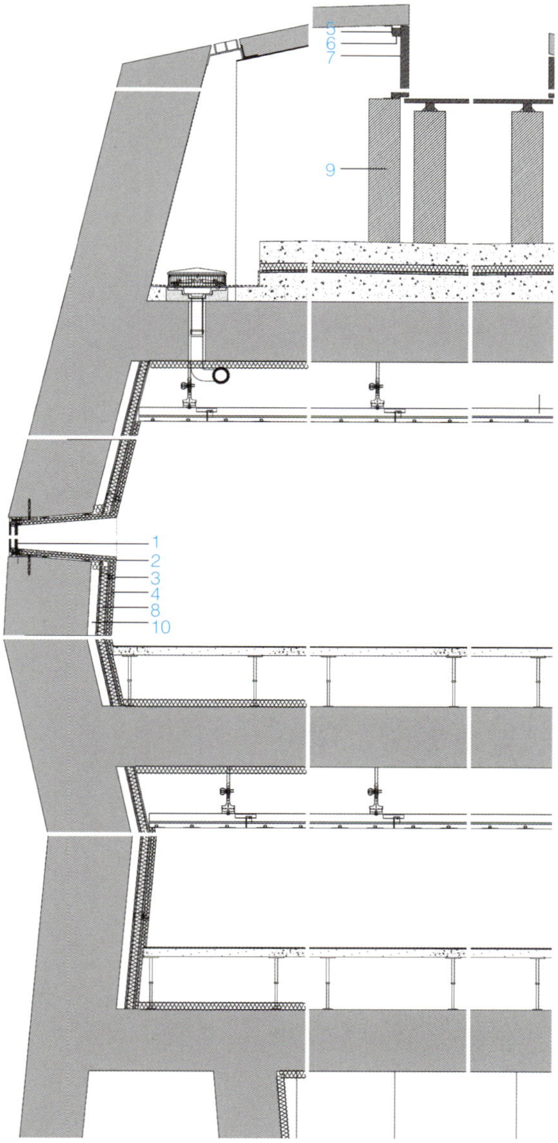

1. Outside windows 1. 外窗
2. Grabs 2. 链钩
3. Metal structure 3. 金属结构
4. Plasterboard 4. 石膏板
5. Metal sheet 5. 金属薄板
6. Ipe wood structure 6. IPE工字钢木结构
7. Ipe slatted wood 7. IPE工字钢板条木结构
8. Thermic insulation 8. 隔热层
9. Concrete wall 9. 混凝土墙
10. Air gap 10. 气隙

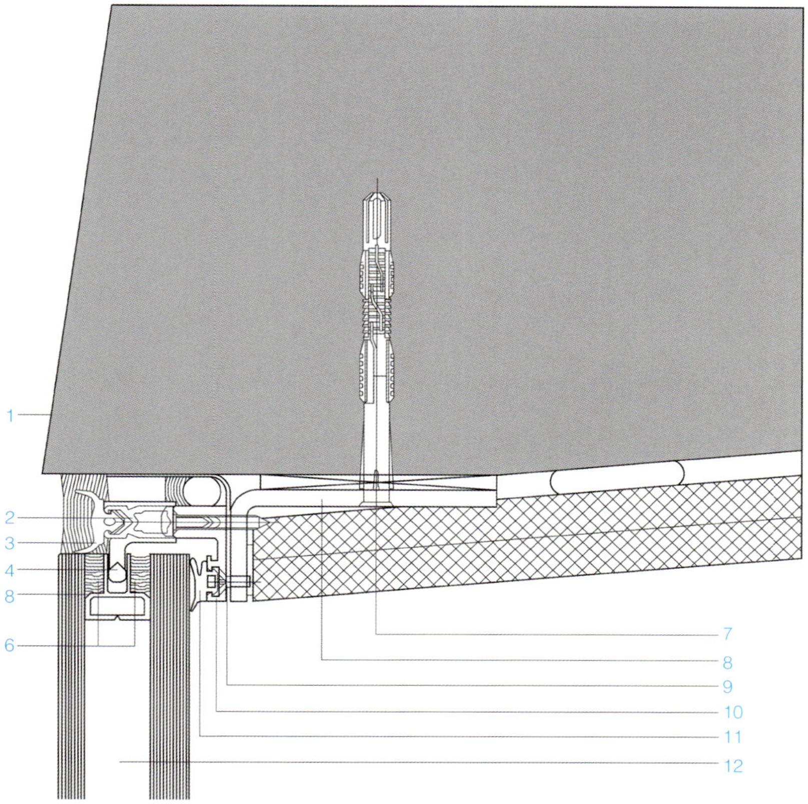

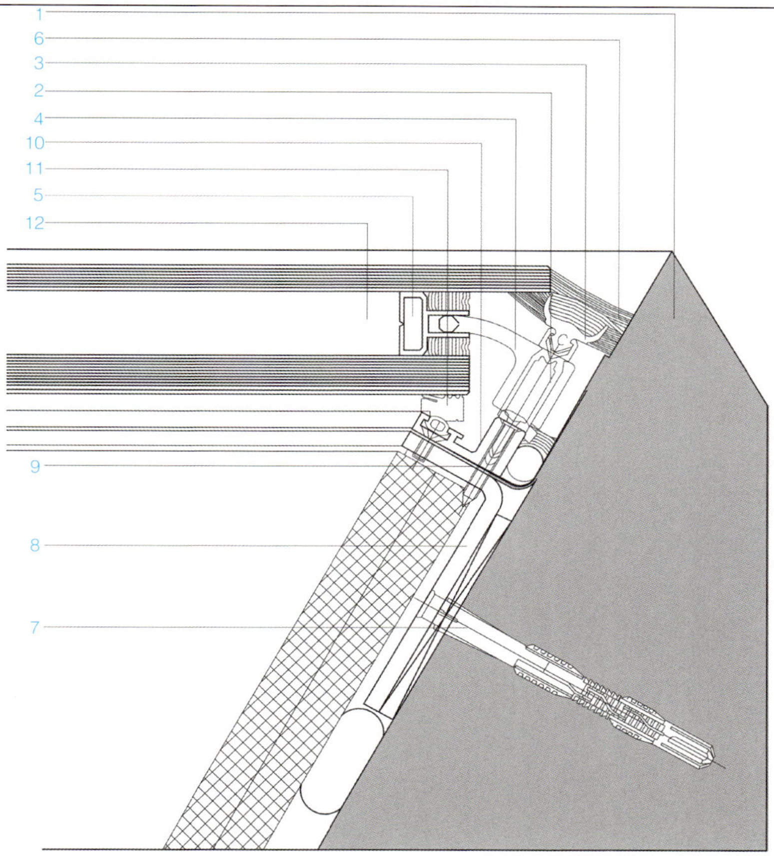

1. White concrete 1. 白色混凝土
2. Insulation 2. 绝缘层
3. Rubber seal 3. 橡胶密封层
4. Glass suport 4. 玻璃支撑结构
5. Interim profile 5. 过渡型材
6. Silicone mastic 6. 有机硅密封层
7. Withdrawl 7. 取料层
8. Stell bar 8. 型钢
9. Tremo layer 9. 涂层
10. Aluminium profile 10. 铝型材
11. Interior glass joint 11. 室内玻璃接缝
12. Double glazing 12. 双层窗户

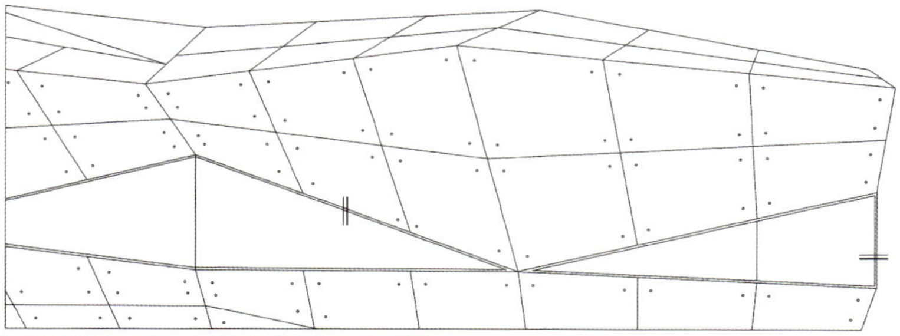

Maasberg Pavilion in Overloon

上洛恩少年管教所

Location: Overloon, The Netherlands
Designers: UArchitects (Eindhoven, The Netherlands)
Photographer: Norbert van Onna
Completion date: 2007

项目地点：荷兰，上洛恩
设计师：UArchitects设计事务所(荷兰，埃因霍温)
摄影师：诺伯特·凡·恩纳
竣工时间：2007年

Award name:
Shortlisted World Architecture Festival Barcelona 2009
1st prize Province Brabant 2009, category Care - The Netherlands
Nominated Bauwelt Prize 2009 - Germany

获奖名称：
2009年巴塞罗那世界建筑节入围；
2009年荷兰布拉班特省关怀类一等奖；
2009年德国Bauwelt奖提名

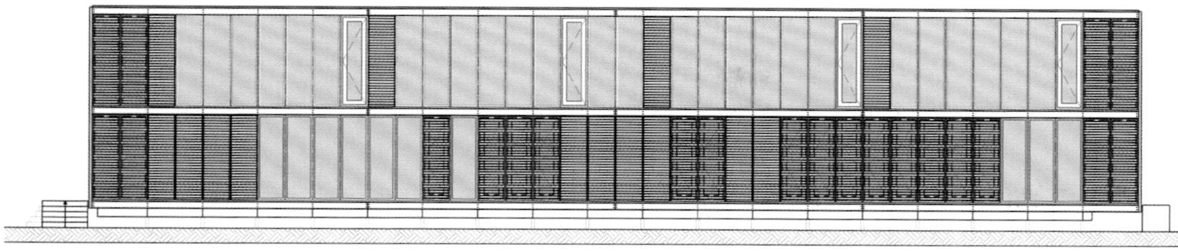

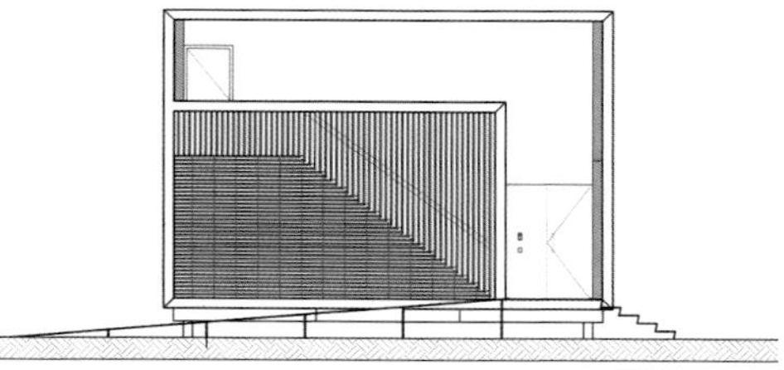

The buildings are renovated and two new living rooms are added. A new pavilion for education, visitors and offices is main part of the master plan.
The client asked for a building which should help the youngsters to return to society, because the youngsters only stay for a while. It's a time-out. The design does not express permanency but the opposite: it is an expression of temporality, like the stay of the youngsters.
The resoluteness of a world focusing on the interior is partly removed. A new and more intensive interaction with the surrounding nature is aimed at the youngsters. The open structure stimulates the daily shifts between living, learning and recreation.
Unit 6 of the "Penitentiaire Inrichting Maashegge" in Overloon is converted from an institution for adults into an autonomously functioning juvenile detention institution. The concept aims at an open, transparent building between the closed prison and the outside world. Visitors can throw a glance at the prison life and the youngsters can look at the outside world, the society in which they will return later on.
For the flexibility at future changes in the program and treatment, the construction is situated in the side façades, the service pipes have been integrated centrally and the side façades are composed of elements which can be changed modularly. The classrooms on the first floor and the visiting room and offices on the ground floor are functioning separately from each other by means of the two outside stairs. At the entrance of the visiting room there is a hall in which both functions are in contact with each other, dramatising the moment of the meeting between the youngsters and the visitors.
By means of the light appearance and the placement on metal feet in the woodland soil, the volume seems to float in the wood. Because of the layered façade in steel, aluminium and Bankirai the building will be merged in the surrounding woodland scenery, showing different shades of grey. The appearance and the rhythmic of the carefully detailed elements of the façade assume a respectful dialogue between environment and the youngsters.
The building looks like a guest in the woodland. As it has not been here for long and was not planning too either. And maybe by its temporality, every idea of time disappears and it will be timeless.

该项目的设计理念是在原有建筑的基础上进行翻新，并增设两个新客厅。构建一个供学习、接待访客和办公于一体的综合型空间是该总体规划的主要部分。
客户要求建筑能够帮助孩子们早日返回社会，这只是他们暂时容身的场所。因此建筑表达了“永远”的反义——短暂，正如孩子们在这里所度过的时间一样。
新规划拆除了部分原有的封闭式内部空间，为青少年营造了一个全新、更具感染力、与周围自然环境完美衔接的空间氛围。开放式结构加强了生活、学习和娱乐之间的互动。
建筑的前身是一所关押成年人的监狱，经过改建后成为青少年管教所。建筑理念是在封闭监狱和外界之间打造一个开放、透明化的空间。访客能够看到孩子们室内生活的场景，同时孩子们也可以随时看到外面的世界——他们即将要返回的地方。
为了使项目能够应变未来的改造，具有较强的灵活性，建筑坐落于侧面，给水管得到集中整合，侧立面成模块化设计。一楼的会客室和办公室被两个户外楼梯分隔开来，二楼是教室。入口处的会见室设有一个大厅，为青少年和访客提供温馨的会面氛围。
浅色外观和金属底座上钢架的设置使整个建筑犹如漂浮在林地之中。由于富有层次感的外立面采用钢材、铝和巴劳木材料，建筑与周围林地景观形成和谐的统一体，并在地面投下点点倒影。独特的建筑外观和充满动感之美的建筑立面力图在环境和青少年之间建立起和谐的对话关系，并使青少年对周围的环境产生尊敬之感。
整个建筑看起来犹如树林中的特约来宾。它不会长期并且也不准备长期存在着。也许正因为短暂，所以每个关于时间的概念都不复存在，从此成为永恒。

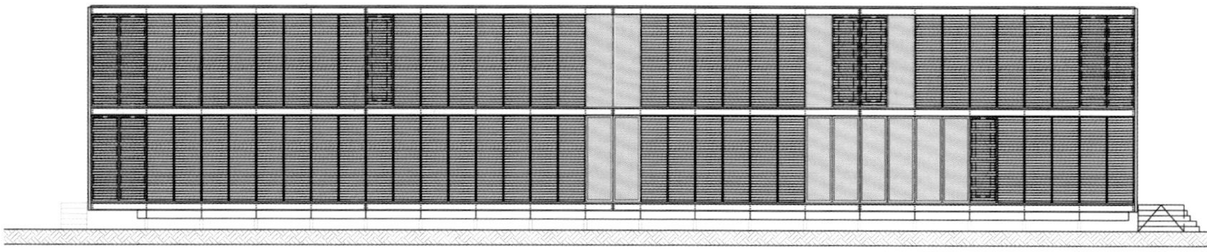

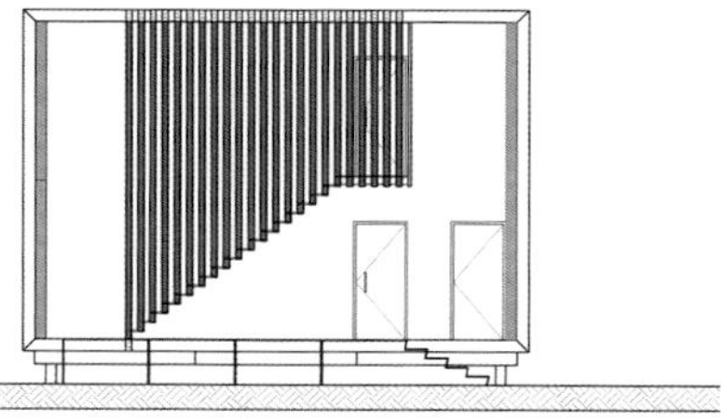

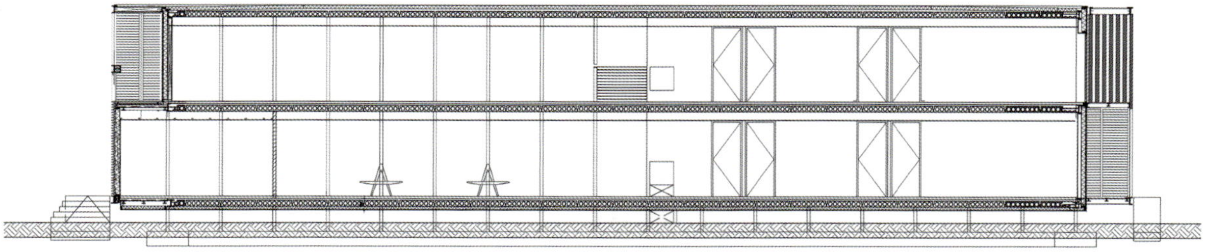

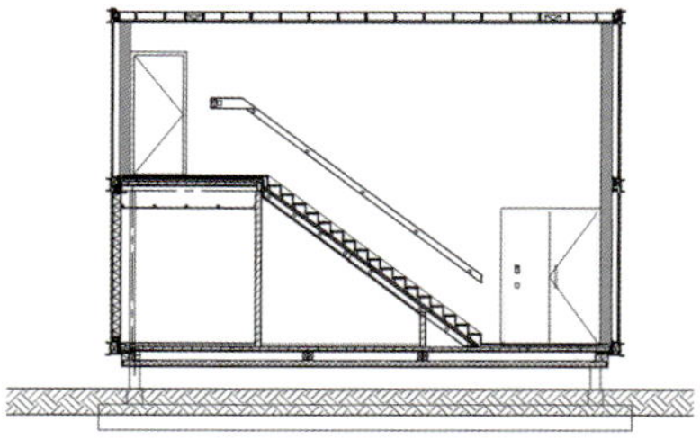

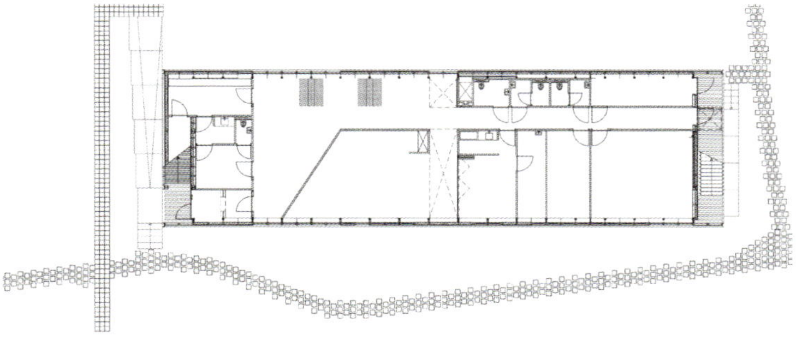

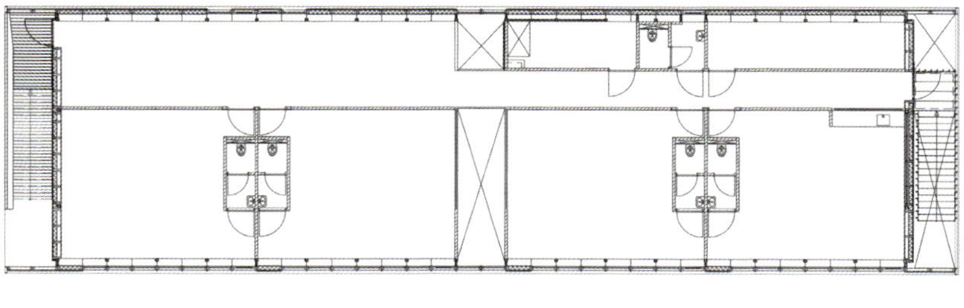

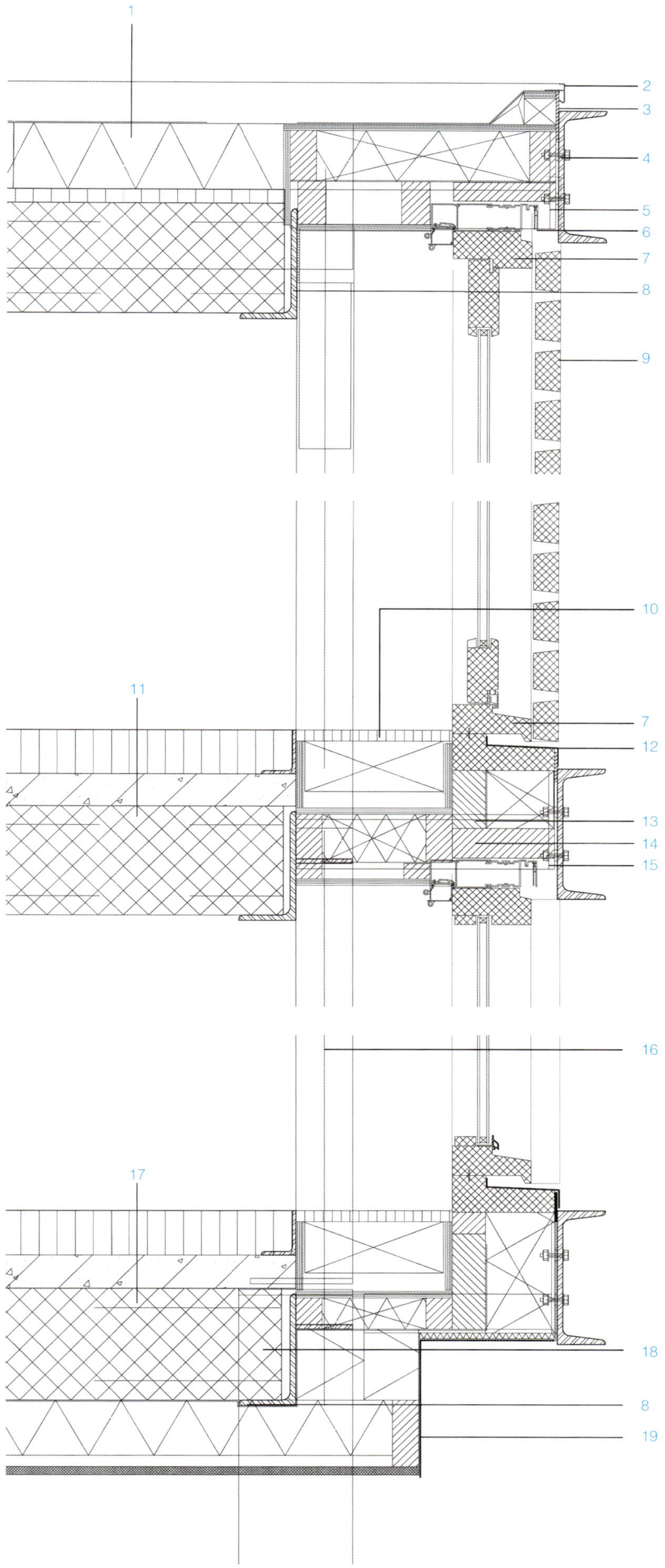

1. Synthetic sealing layer; 120mm rigid foam insulation; 200mm prefab Concrete floor
2. Aluminium roof covering
3. 5 mm steel strip welded to unp
4. Steel section unp 240, thermal interruption with plastic strip 5mm
5. 2 X 70/70/7mm steel I-section
6. Ventilation grating
7. 67/139mm meranti window-frame, painted
8. Steel I-section, 200/100/10mm
9. 75/44mm wooden bankiray latticework, 90mm apart on steel framework
10. Aluminium grating over heating duct
11. 80Mm concrete finishing layer; 60mm concrete distribution layer; 200mm prefab concrete floor
12. Aluminium
13. 59/171mm wooden beam
14. 54/180mm wooden beam
15. 2 X 80/80/5mm steel I-section
16. 100/100/10mm steel column (in view)
17. 80mm concrete finishing layer; 60mm concrete distribution layer; 200mm prefab concrete floor; 140mm insulation; weather resistant fibreboard on wooden framework
18. Steel section unp 220
19. Galvanised steel, dark-grey

1. 合成密封层；120毫米的硬质泡沫绝缘层；200毫米的预制混凝土楼板
2. 铝屋顶盖板
3. 与槽钢焊接的5毫米钢带
4. 槽钢 240，采用5毫米的塑料带进行隔热
5. 2 ×70/70/7毫米L形钢材剖面
6. 通风格栅
7. 67/139毫米着色梅兰蒂木窗框
8. L形轴压钢柱，200/100/10毫米
9. 75/44毫米木格栅，距离钢框架90毫米
10. 加热管上的铝制格栅
11. 80毫米的混凝土层；60毫米的分布层；200毫米的预制混凝土楼板
12. 铝
13. 59/171毫米木梁
14. 54/180毫米木梁
15. 2 ×80/80/5毫米L形轴压钢柱
16. 100/100/10毫米钢柱（目测）
17. 80毫米的混凝土层；60毫米的分布层；200毫米的预制混凝土楼板，140毫米的绝缘层；耐久纤维板木框架
18. 槽钢220
19. 暗灰色镀锌钢管

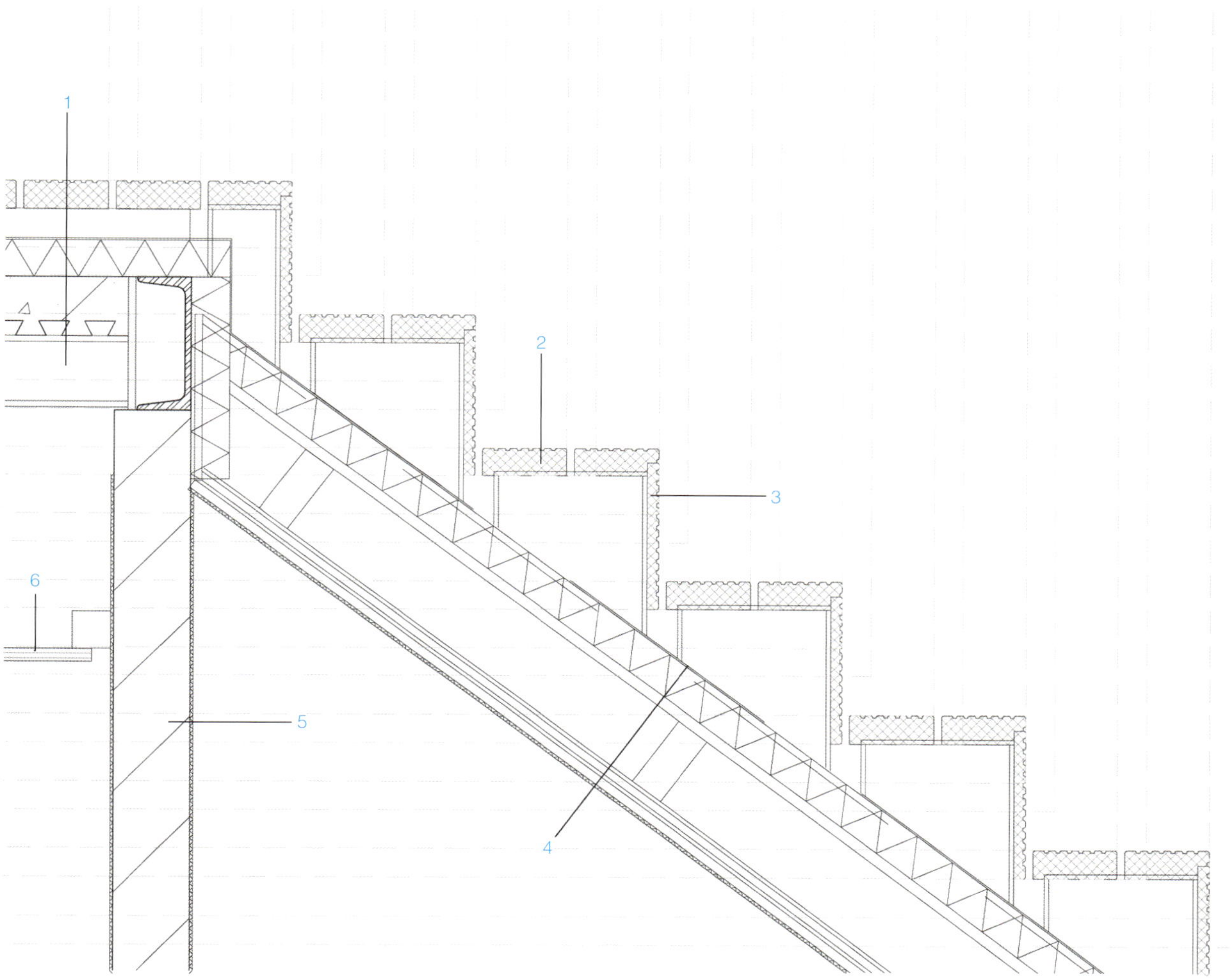

1. 110/38mm bankiray parts, 120mm apart screwed on underlying bankiray parts; sealing layer, PUR insulation; 80mm concrete floor on profiled steel plating; steel section HE100A
2. 110/38mm bankiray parts, 120mm apart on steel handles
3. 15x203mm bankiray risers on steel handles
4. Sealing layer; PUR insulation; 19mm underlayment; wooden joisting between steel sections UNP 200; sealing layer; latticework; 12,5mm fibreboard; plaster
5. Lime-sand brick wall
6. 15mm plywood ceiling

1. 110/38毫米 bankiray木料零件，120毫米间距，固定在下面的bankiray木料零件之上；密封层，聚氨酯绝缘层；80毫米的镀钢异型混凝土楼板，H型钢100A
2. 110/38毫米 bankiray木料零件，120毫米间距，位于钢柄之上
3. 钢柄上的15x203毫米 bankiray木料升管
4. 密封层，聚氨酯绝缘，19毫米的垫层；型钢UNP200之间的木格栅；密封层；格栅；12.5毫米的纤维板；石膏板
5. 灰砂砖壁
6. 15毫米的胶合板天花板

Head Office Irdeto

爱迪德公司总部

Location: Hoofddorp, The Netherlands
Designers: MIII Architects
Photographer: John Lewis Marshall
Completion date: 2010

项目地点：荷兰，霍夫多夫
设计师：MIII建筑师事务所
摄影师：约翰·刘易斯·马歇尔设计工作室
竣工时间：2010年

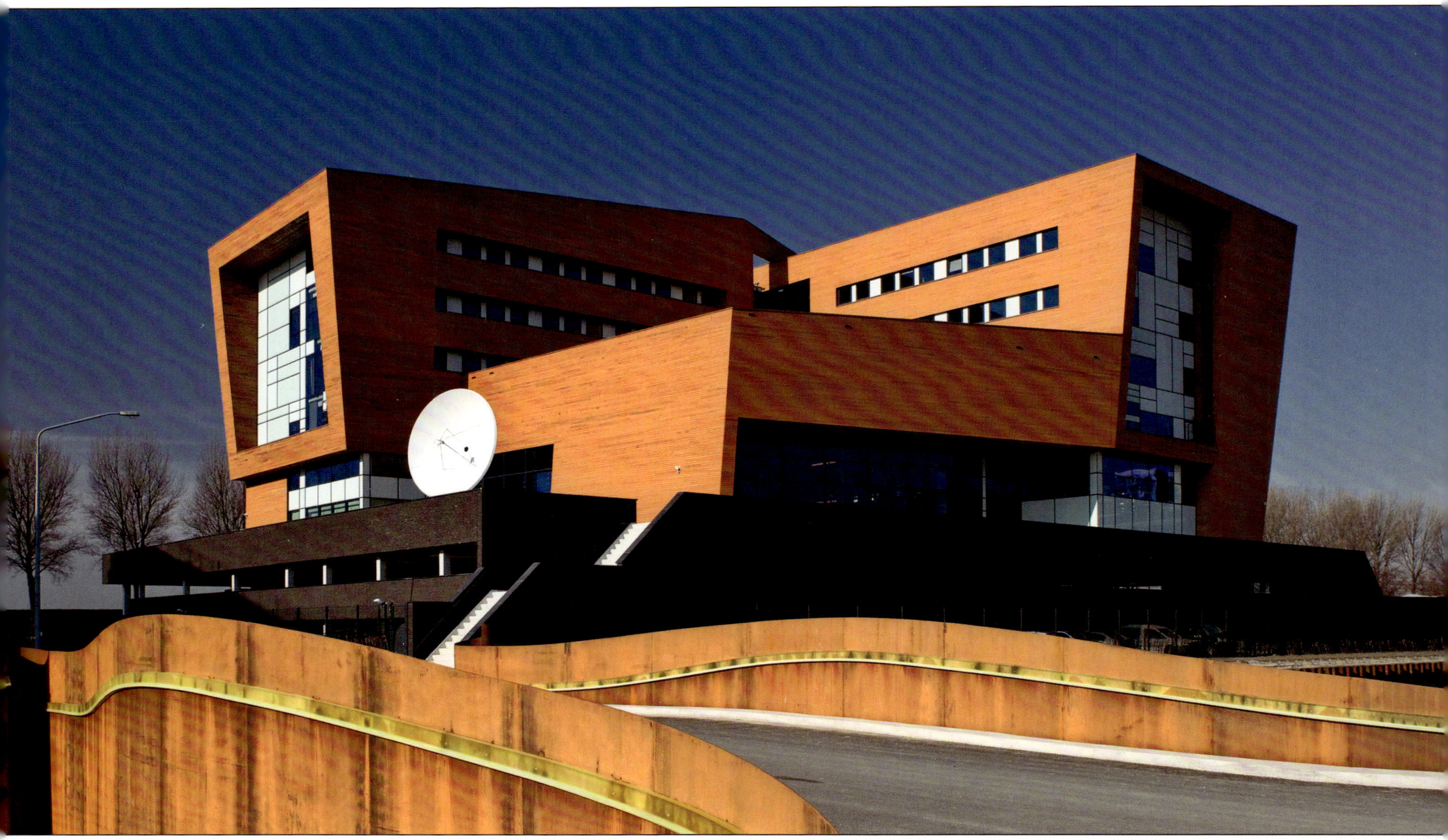

Irdeto, the worldwide market leader in technology for content security of digital TV, widescreen TV and mobile TV, has recently moved into their new head office in Hoofddorp, The Netherlands. This completely sustainable 10,000 square metres office building was designed by MIII Architects and meets the GreenCalc-score B. No surprise that the core values of developing this building were: sustainability, flexibility and future orientation.

The charcoal-coloured brick pedestal elevates the building and can provide 200 vehicles of a two-storey-high parking space and enters a loading dock and a warehouse facility. The building is sculpted in three wings each of six storey's high, around a central located atrium. Undeniable eye catcher is the wooden curtain wall that covers the concrete construction and knots the three volumes to one building. It's not very common to get an approval from an investor for using wood for an industrial building and it took MIII great effort to convince them. The decisive reason was in MIII's earlier experience that it constructed properly and treated with the right materials, wood will remain well. During this process MIII Architects and the investor chosed and treated Western Red Ceder in a vivid terra-cotta coloured finish.

Inside the Atrium the tree building volumes are highly visible because of the repetitive use of the Western Red Ceder so the volumes can physically continue in the interior. Another aspect of the interior is the amount of large open spaces. Acoustical and visible shut spaces are a "necessary evil" in an office building but in Irdeto's office they're in a model that mostly creates corners instead of dividing the floor in separate chambers. The view from some of the storey-high windows do not only show the surrounding but the interesting tension between the intimate character of wood and its engineered metal construction as well.

爱迪德公司是全球数字电视、宽屏电视和移动电视内容安全技术的市场先锋，刚刚迁入到荷兰霍夫多夫市的新总部之中。这个占地10,000平方米的办公楼由MIII建筑师事务所设计，符合GreenCalc-Score绿色环保评分标准。该建筑设计的核心理念是打造集可持续性、灵活性和未来适应性于一体的办公空间。

由黑色砖石筑成的基座将建筑缓缓抬起，并形成了一个两层高的停车空间、一个装货码头和一个仓库。建筑共有三个侧楼，每一个侧楼均有6层高，围绕一个中央中庭展开布局。覆盖整个混凝土结构、并将三个体量衔接为一体的木质幕墙十分引人注目。对于一个工业建筑来说，说服投资方采用木材作为主要建筑材料无疑是一件难事，而MIII建筑师事务所用实际行动证明了其可行性，其决定性的原因在于MIII建筑师事务所早期的建设经验向人们证明了如果处理得当，木料可以扮演良好的角色。在该项目的设计过程中，MIII建筑师事务所以及投资方最终决定采用西部红松为原料，并在其外表涂以棕橙色。

在中庭内部，因建筑采用了西部红松，建筑的体量显得尤为抢眼，并与建筑的外观相得益彰。室内另一个亮点之处在于大型开放空间。在普通的办公建筑之中，隔音可视封闭空间经常被视为房间划分的“必要之恶”，然而在爱迪德公司总部中，它们巧妙地构建出若干角落，从而使空间分隔开来。从高层的窗口处不仅能够将周边的景致尽收眼底，同时可以领略到木料与其工程金属结构之间的张力。

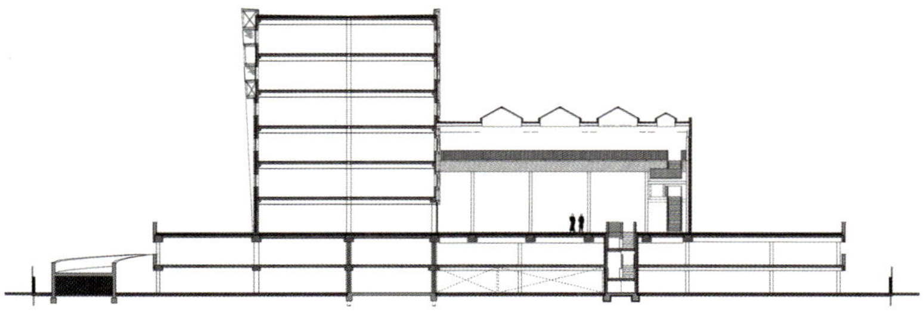

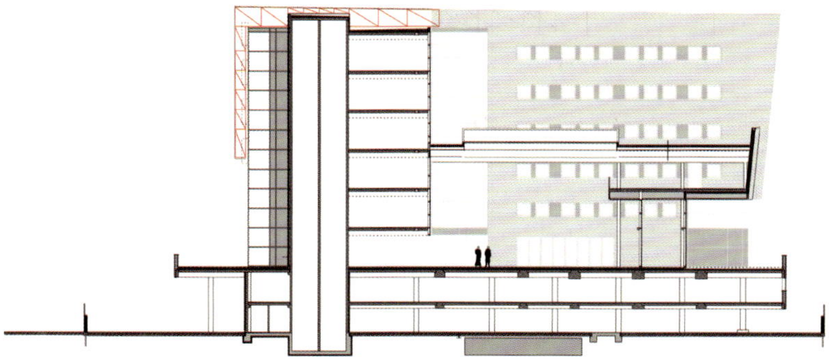

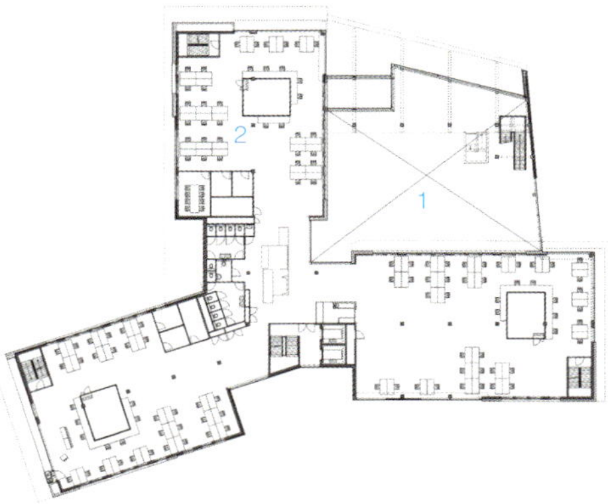

1. Atrium
2. Office area

1. 中庭
2. 办公区

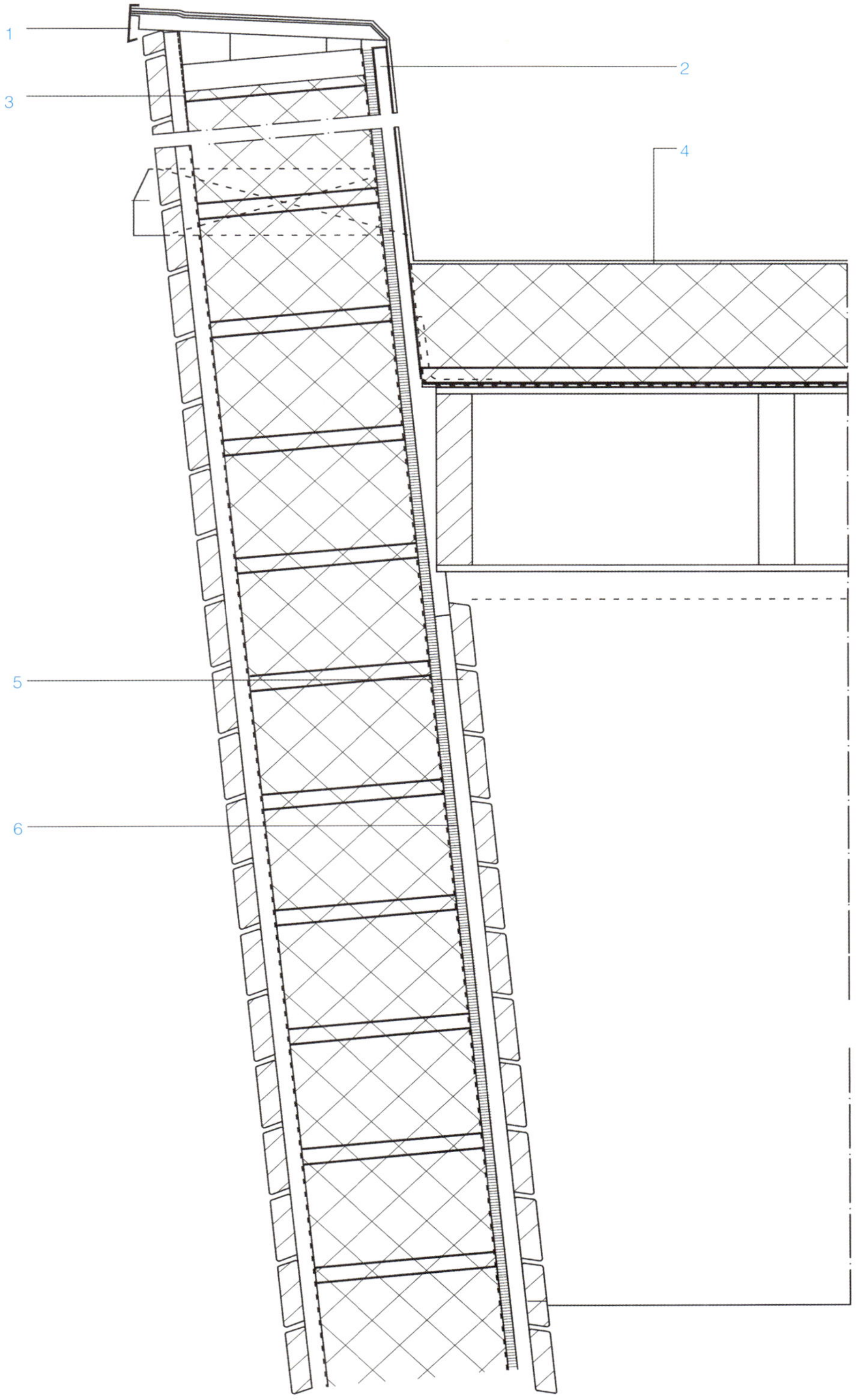

1. Aluminium edging strip
2. Plywood edge of the roof timbering 18mm
3. Damp foil
4. Polymeric roof covering
5. Treated pinewood laths on battens (material as pinewood laths)
6. Timberframe construction-dampproof foil-mineral wool-wooden batten 38mmx235mm-damp resistant foil fermacel 12.5mm

1. 铝制封边条
2. 屋顶木结构胶合板切边18毫米
3. 阻尼金属薄片
4. 高分子屋顶覆盖
5. 处理木板条（如松木板条材料）
6. 木材框架结构，防潮铝箔矿棉，木条板 38x235毫米+防潮天花板12.5毫米

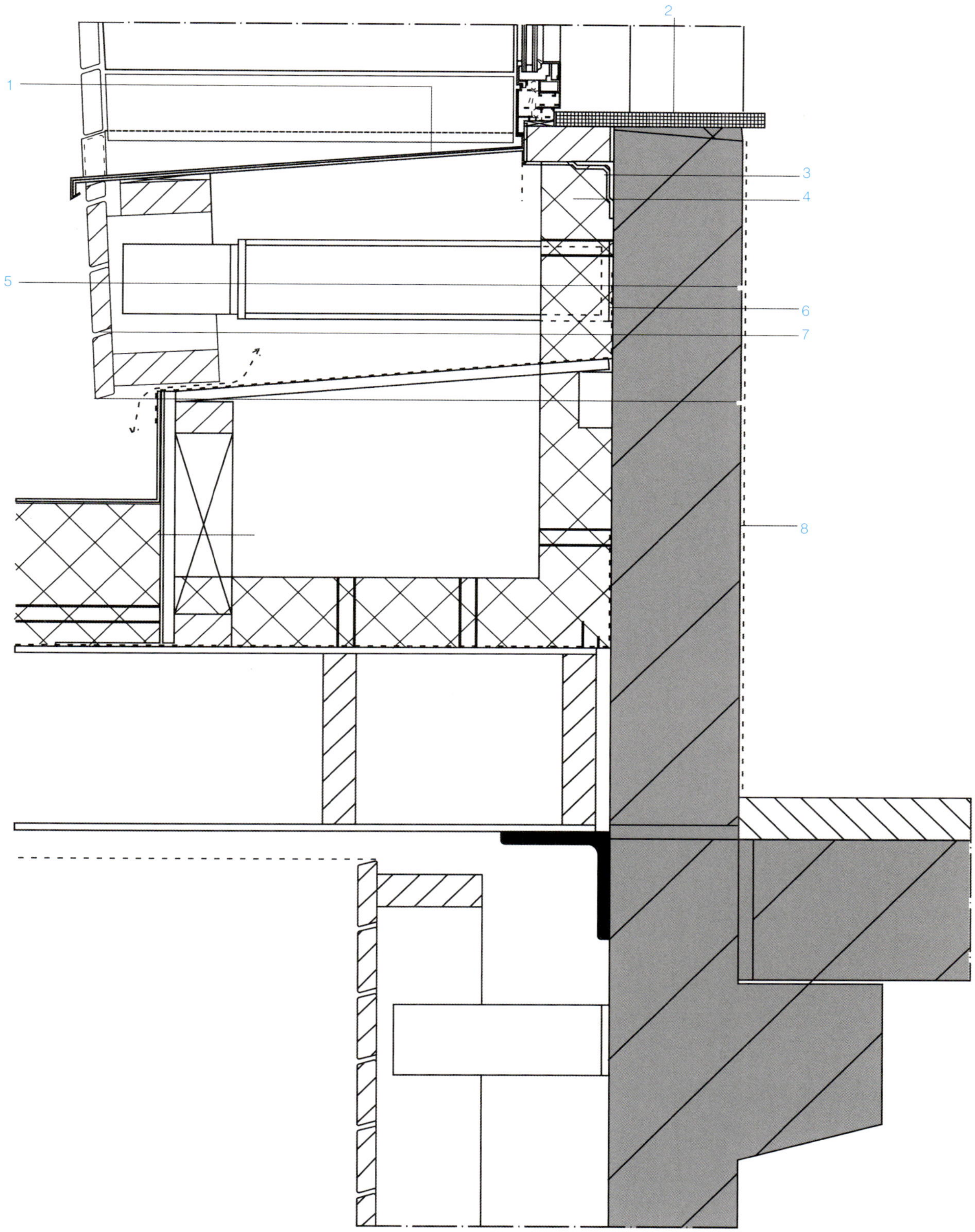

1. Aluminium water hammer with foil and bulkheads
2. Window sill nibo granite black 20mm
3. Hardcast
4. Glass wool isolation isover 433 plus ultra isolationboard with aluminium foil /glass fibre
5. Steel L-stirrup for stening prefab frame structural engineer
6. Neoprene
7. Façad (out/in): treated pinewood battens prefab wooden frame
8. Spack

1. 防水壁中设置的铝制水锤吸纳器与金属薄片
2. 窗台板——黑色花岗岩20毫米
3. 硬铸铁
4. 玻璃丝绝缘层与铝箔/玻璃纤维构成的绝缘板
5. 用于固定预制框架结构的L形钢
6. 氯丁橡胶
7. 立面（外部/内部）：处理松木板条预制木制框架
8. 外包层

Office Building VDAB

VDAB办公楼

Location: St-Niklaas, Belgium
Designers: BOB361 Architecten
Photographers: Nullens André, Frederik Vercruysse
Completion date: 2005

项目地点：比利时，圣尼可拉斯
设计师：BOB361建筑师事务所
摄影师：那伦斯·安德烈，弗雷德里克·沃克卢斯
竣工时间：2005年

Award name:
2009 Shortlisted, World Architecture Festival

获奖名称：
2009年世界建筑节入围奖

Considering the urban anchoring of the site, the angles of this building block are the regulating elements. The VDAB-project anchors itself with the existing context on the corner of the Noordlaan-Drie Koningenstraat, which is in direct relation with various public functions (post, ministry of finances), located near the main road.
Concerning the implantation of the building, the volume was entirely concentrated on the street side. This offers several advantages in the field of compactness, economy and energy assessment, acoustic comfort for the offices on the garden side, large distances to the neighbouring houses and avoids fragmentation of the open green space. For the acoustic comfort on the street side, patios were added. As a result, a simple image of the emplacement between the built and green space appears.
Because of the excavating cost the lightly polluted soil, the actual volume to be removed was minimised by mooring the parking into a slope. The remaining volume of soil extracted was used to form a landscaped garden and is naturally purified by the use of fotoremediating plants. The open and slightly sloped car park, offers thus, a pleasant solution that is naturally lit and ventilated, with clear views on the garden and other open spaces. Furthermore, the principal interior distribution passage way of the building, runs from the ground level to the first floor along a slope that is parallel to the parking space; This walkway links the public spaces on the ground floor in a natural and spatial manner with semi-public spaces on the higher levels.

考虑到所在地的特殊地理位置，建筑的外观呈多角度设计，与周围的环境以及交通要道附近的Noordlaan–Drie Koningenstraat区（邮政、财政部等所在地）自然融为一体。
办公大楼的体量面向街道而设。该设计充分考虑了紧凑简洁、经济节能的因素，另外，还要确保面向花园的办公空间部分保持静谧、宜人的环境，与邻近民居保持一定距离，以及防止对外部绿色空间的破坏。出于噪音处理的考虑，空间内还增设了天井。优雅的建筑与绿意盎然的自然环境遥遥相望、互相映衬。
由于悬空停车场建立在山坡之上，挖掘出的泥土部分被用到花园之中，巧妙地打造了一个景观园林；而园中植物则能够帮助这些在建造过程中被轻度污染的土壤在日后得到逐渐的净化。开放、略带缓坡的停车场内光线充足、通风性能较好，能够将花园和其他开放空间的景致尽收眼底。室内主要通道沿着停车场的斜坡从建筑的底层一直通往二层空间，并自然地将底层的公共区与楼上的半公共区有机衔接在一起。

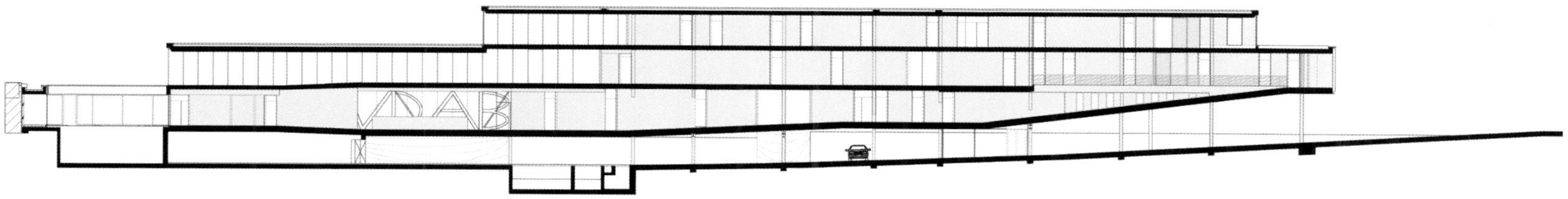

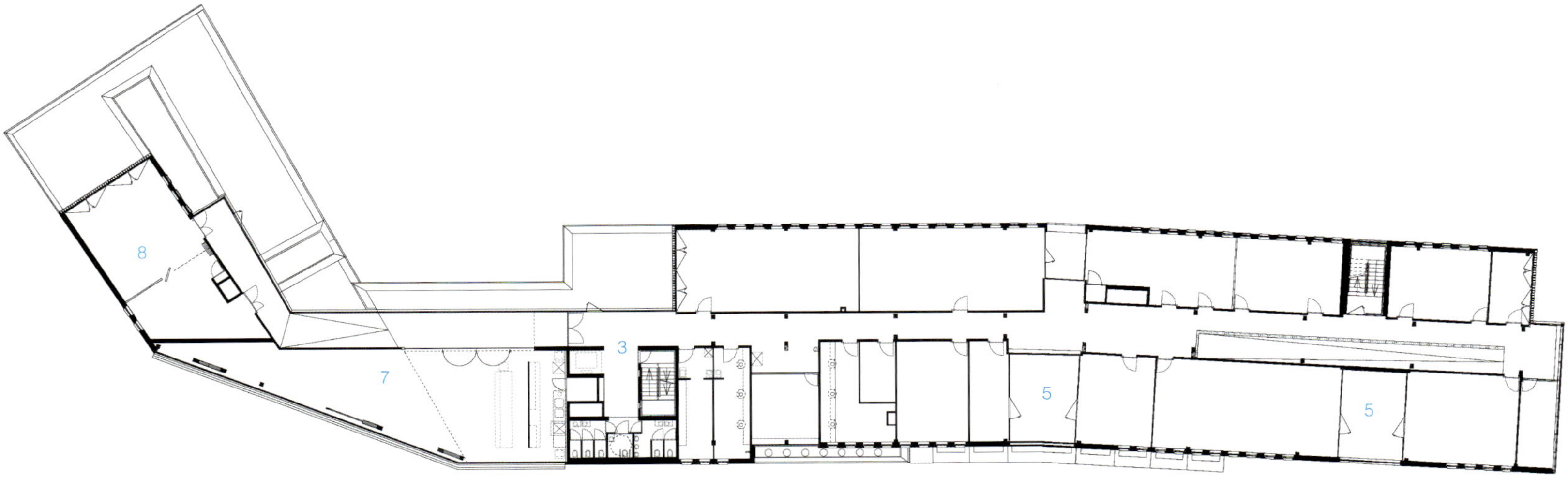

1. Covered outside space
2. Entrance hall
3. Pedestrian slope
4. Hall
5. Patio
6. Parking space
7. Cafetaria
8. Meeting room

1. 加盖外部空间
2. 门厅
3. 行人坡道
4. 大厅
5. 天井
6. 停车空间
7. 自助餐厅
8. 会议室

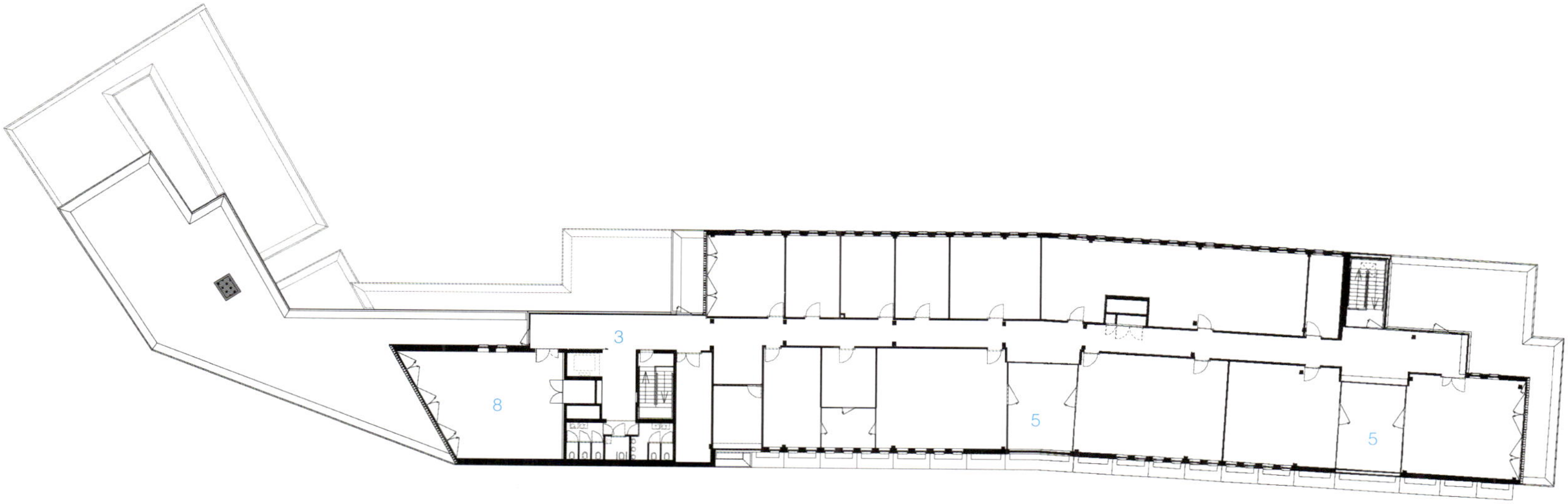
3
8
5
5

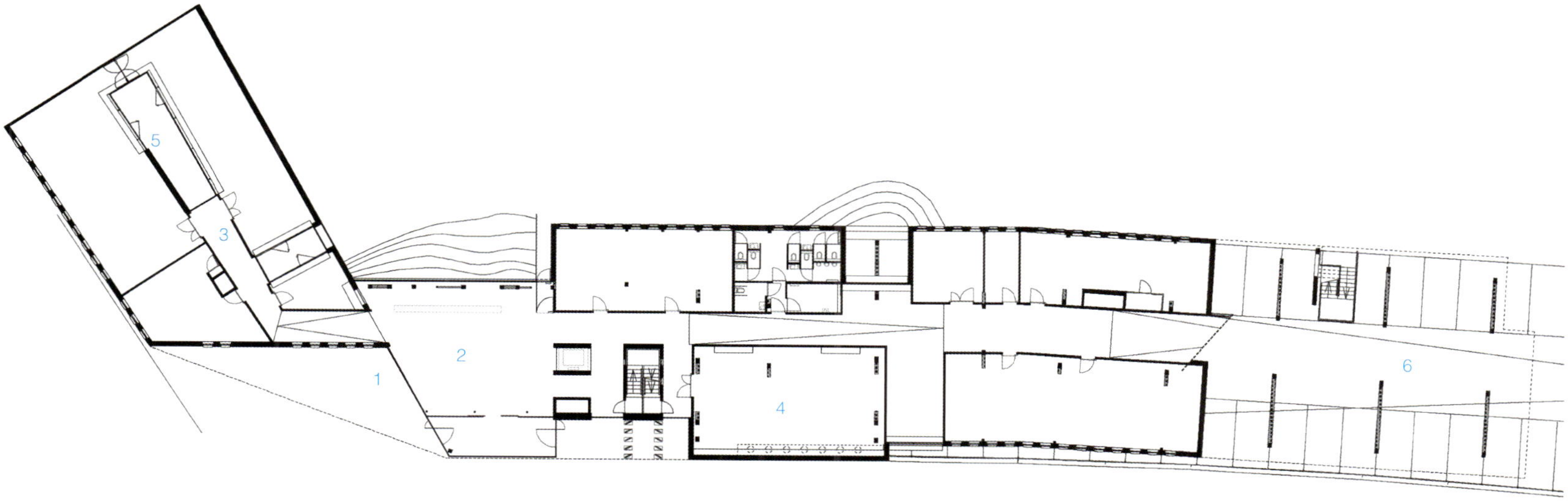
5
3
2
1
4
6

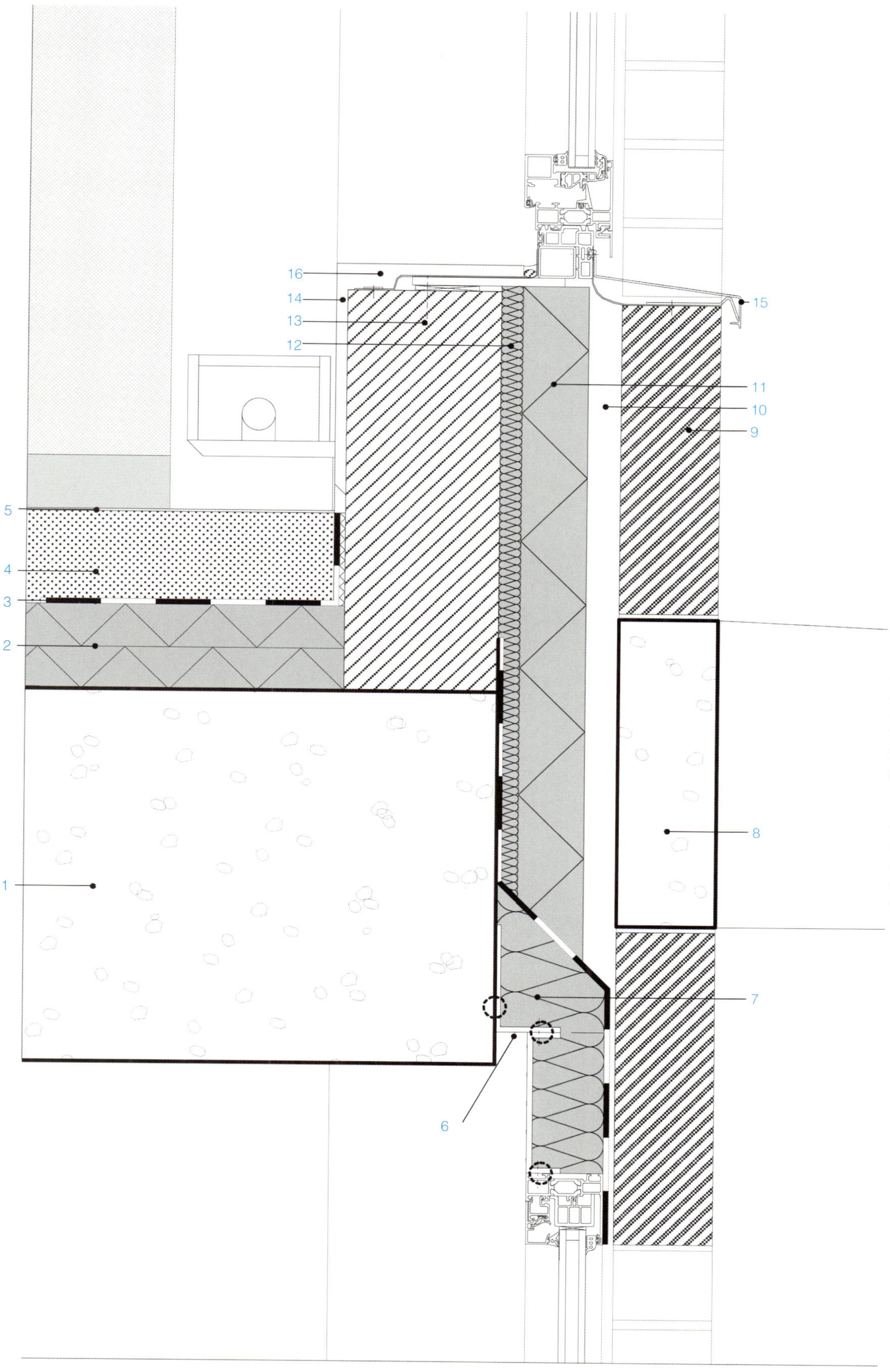

1. Concrete slab 35cm
2. Pir insulation 2cmx4cm
3. Polyethylene waterproof membrane
4. Screed 9cm
5. Polyurethane floor
6. Sealant strip
7. Thermal insulation
8. Precast concrete
9. Brick 9cm
10. Ventilated cavity
11. Pir thermal insulation 6cm
12. Mineral wool thermal insulation 2cm
13. Load-bearing brickwork
14. Plaster 1cm
15. Aluminium window sill
16. Plaster

1. 35厘米厚混凝土板
2. 2x4厘米红外绝缘板
3. 聚乙烯防水卷材
4. 9厘米砂浆底层
5. 聚氨酯楼板
6. 密封带
7. 保温
8. 预制混凝土
9. 9厘米砖
10. 通风腔口
11. 6厘米红外保温板
12. 2厘米岩棉保温绝缘板
13. 承重砖
14. 石膏厚为1厘米
15. 铝合金窗台
16. 石膏

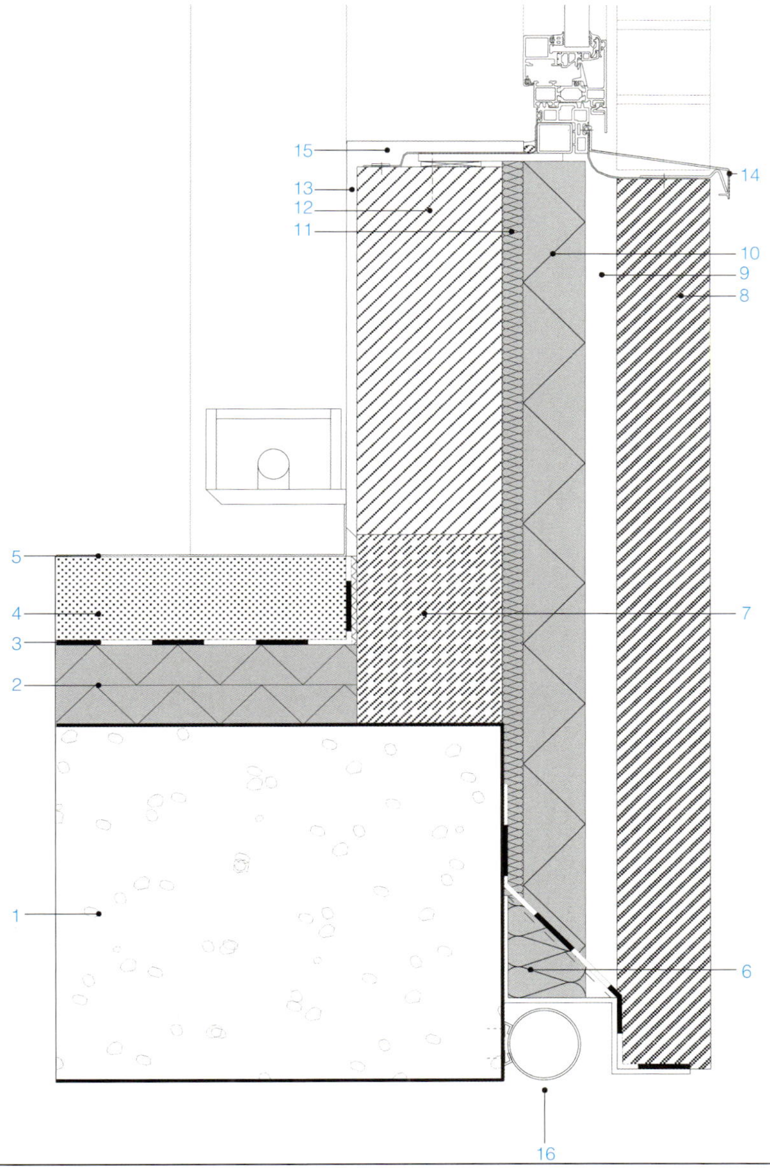

1. Concrete slab 35cm
2. Pir insulation 2cm x 4cm
3. Polyethylene waterproof membrane
4. Screed 9cm
5. Polyurethane floor
6. Thermal insulation
7. Precast concrete
8. Brick 9cm
9. Ventilated cavity
10. Pir thermal insulation 6cm
11. Mineral wool thermal insulation 2cm
12. Load-bearing brickwork
13. Plaster 1cm
14. Aluminium window sill
15. Plaster
16. Lighting

1. 35厘米厚混凝土板
2. 2x4厘米红外绝缘板
3. 聚乙烯防水卷材
4. 9厘米砂浆底层
5. 聚氨酯楼板
6. 保温
7. 预制混凝土
8. 9厘米砖
9. 通风腔口
10. 6厘米红外保温板
11. 2厘米岩棉保温绝缘板
12. 承重砖
13. 石膏厚为1厘米
14. 铝合金窗台
15. 石膏
16. 照明

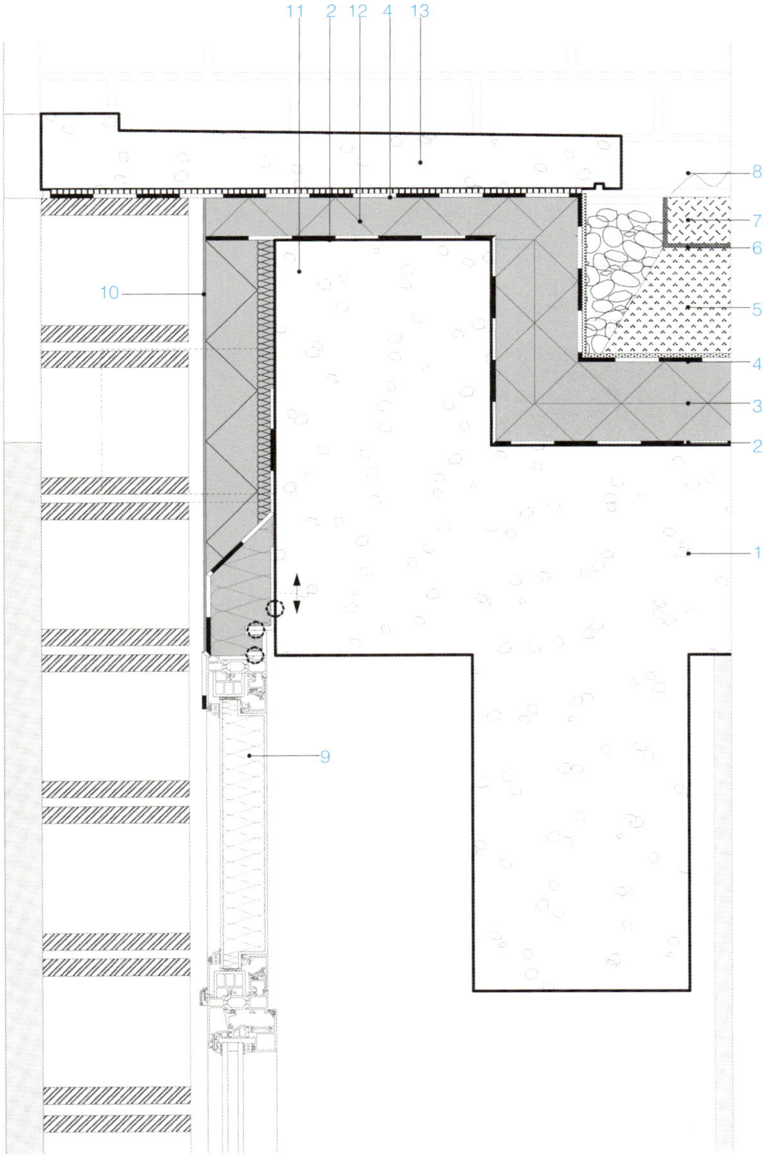

1. Concrete slab 25cm
2. Waterproof membrane
3. PIR insulation 2x 5cm
4. Flashing
5. Drainage 10cm
6. Filter
7. Soil
8. Vegetation
9. Steel sandwich panel filled with mineral wool
10. Vapour open membrane, black
11. Concrete beam
12. Pir insulation 5cm
13. Concrete capping

1. 25厘米厚混凝土板
2. 防水卷材
3. 2x 5厘米红外绝缘板
4. 防雨板
5. 10厘米排水系统
6. 滤波器
7. 土壤
8. 植被
9. 夹芯板与矿棉填充
10. 黑色蒸汽膜
11. 混凝土梁
12. 5厘米厚红外绝缘板
13. 混凝土封顶

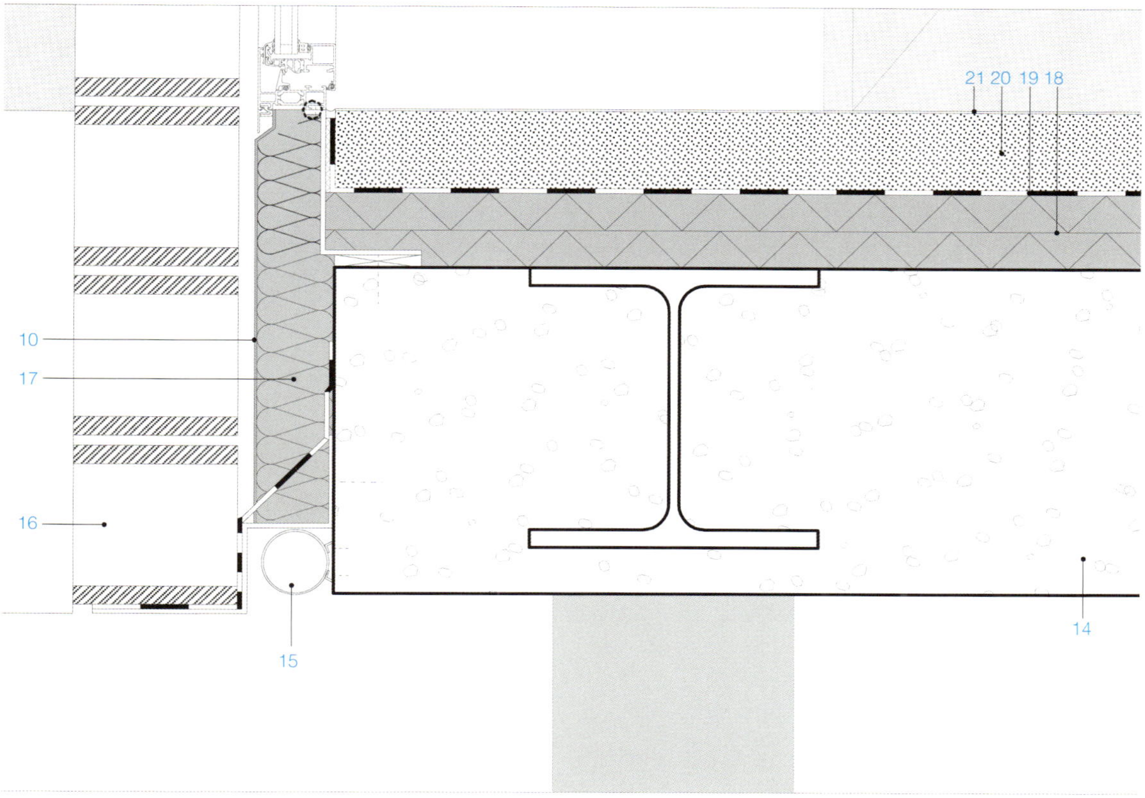

10. Vapour open membrane, black
14. Concrete slab 35cm
15. Lighting
16. Claustra, horizontally and vertically reinforced
17. Mineral wool thermal insulation
18. Pir insulation 2cmx4cm
19. Polyethylene waterproof membrane
20. Screed 9cm
21. Polyurethane floor

10. 黑色蒸汽膜
14. 混凝土板厚度为35厘米
15. 照明
16. 屏状核，横向和纵向加固
17. 岩棉绝缘层
18. 2x4厘米红外绝缘层
19. 聚乙烯防水卷材
20. 9厘米厚砂浆底层
21. 聚氨酯地板

Horizontal Skyscraper - Vanke Centre

水平摩天楼——万科中心

Location: Shenzhen, China
Designers: Steven Holl Architects
Photographer: Iwan Baan and Shu He
Completion date: 2009

项目地点：中国，深圳
设计师：史蒂芬·霍尔建筑师事务所
摄影师：伊万·班，舒赫
竣工时间：2009年

Award name:
2009 AIA New York Honour Award
2010 AIA Institute Honour Award

获奖名称：
2009年美国建筑师协会纽约荣誉奖
2010年美国建筑师协会荣誉奖

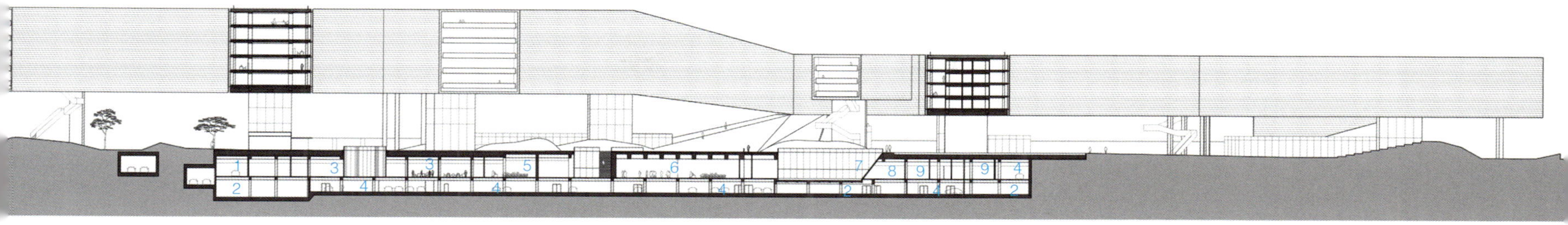

The Vanke Centre is one of the first, highest rated USGBC, LEED Platinum Certified Project in China. Renewable materials were used in Vanke Headquarter's wing. Bamboo, this highly renewable material, which is easily available in China, was used for doors, floors and furniture throughout the Vanke Headquarters instead of using raw materials or exotic woods. Green carpet the InterfaceFLOR Carpet tiles were used throughout the open office area. This carpet is a cradle-to-cradle product, meaning that it is not only produced from recycled materials, but that the manufacturer agrees to collect any damaged carpet and to recycle it into other carpet or products.

The Vanke Headquarters used Greenscreen solar shading fabrics from Nysan – a PVC free product that contains no VOCs (Volatile Organic Compounds). Not only does the product not "off-gas" during its life time, but it is also easier and quicker to recycle and divert to landfills.

The project is both a building and a landscape, a delicate intertwining of sophisticated engineering and the natural environment. By raising the building off of the ground plane, an open, publicly accessible park creates new social space in an otherwise closed and privatised community.

The site area is approximately 60,000 square metres: of which 45,000 square metres was planted. With the addition of the planted roof area of the main building (approximately 15,000 square metres) - the total planted area of the project is roughly equal to the site before development.

万科中心将成为首个通过美国绿色建筑委员会LEED白金认证的中国项目。建筑内部的门、地板、家具等全部采用了中国国内的常见材料——可再生竹材，代替了其他原材料和异地木料。环保地毯——由英特飞公司生产的方块毯铺满了整个开放式办公区。该地毯属于“从摇篮到摇篮”产品，这就意味着该拼毯不仅由再生材料制成，同时也意味着制造商将损坏的地毯回收之后可将其重新利用，使之成为其他地毯和产品。

万科总部采用不含挥发性有机化合物的产品作为遮阳系统的材料。该产品在使用期间不产生任何有毒气体，同时易回收、处理。

该建筑外观优美、高雅，是城市中一道亮丽的建筑景观。精致独特的外观与周围自然环境巧妙地融合在一起。架空的建筑底部形成对流通风良好的微气候，吸引着人们在此驻足休憩。

建筑的总面积约为60,000平方米，占地面积为45,000平方米，加上主楼（约占地15,000平方米）的绿化屋顶面积，整个大楼的建筑面积与建筑所在地的原面积相近。

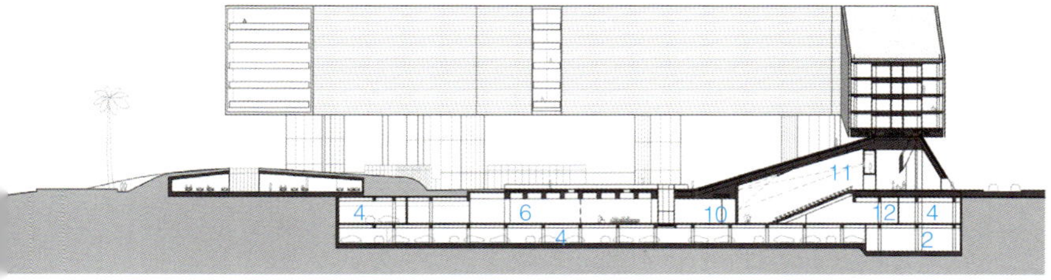

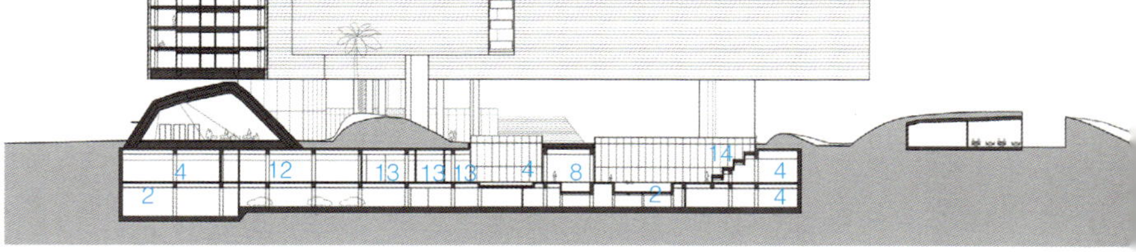

1. Service
2. Mechanic room
3. Meeting room
4. Parking
5. Video conference room
6. Conference centre
7. Water garden
8. Interior swimming pool
9. Locker room
10. Conference
11. Auditorium
12. Hotel servie
13. SPA
14. Outdoor swimming pool

1. 服务区
2. 机械室
3. 会议室
4. 停车场
5. 视讯会议
6. 会议中心
7. 水景花园
8. 室内游泳池
9. 更衣室
10. 会议中心门厅
11. 观众厅
12. 酒店服务
13. 温泉洗浴
14. 室外游泳池

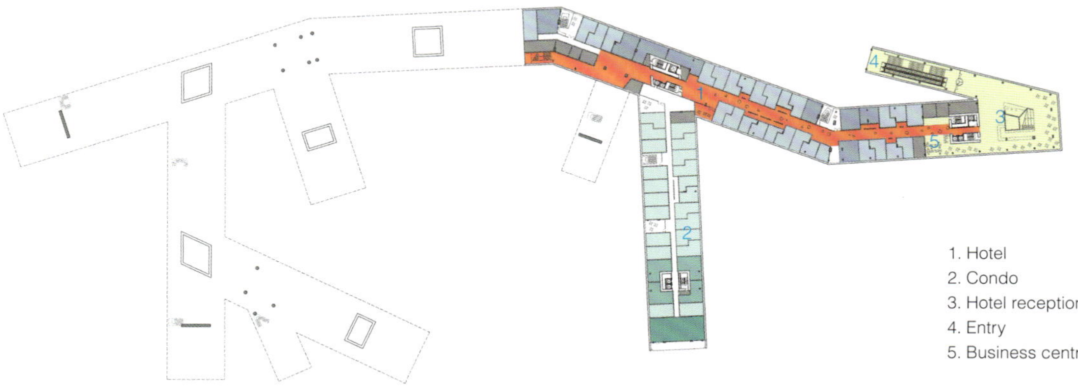

1. Hotel
2. Condo
3. Hotel reception
4. Entry
5. Business centre

1. 酒店
2. 产权式酒店
3. 酒店大堂
4. 入口
5. 商务中心

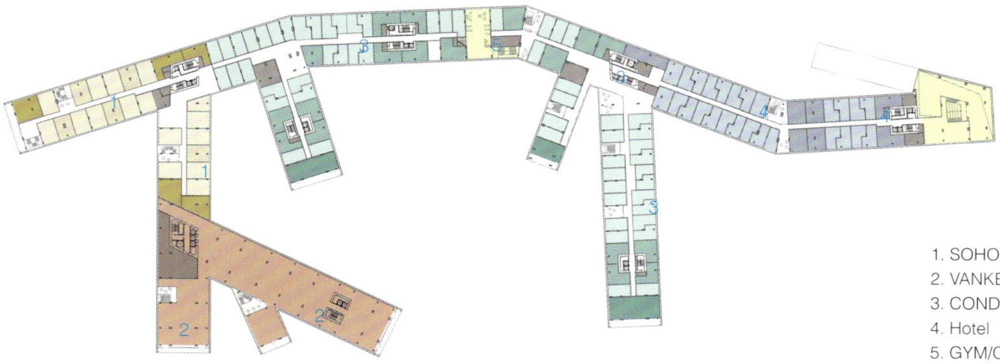

1. SOHO office
2. VANKE centre
3. CONDO
4. Hotel
5. GYM/Community space

1. 家庭办公
2. 万科中心
3. 产权式酒店
4. 酒店
5. 健身房/活动室

1. Floor finish
2. Aluminium panel shadow box
3. Fire insulation infill
4. Fire separation laminated glass
5. Obscure glass
6. Exposed concrete
7. Exterior glass with steel back-up behind
8. Operable window
9. Clear glass

1. 楼板完成面
2. 铝制遮阳玻璃后加保温装置
3. 防火绝缘材料填充
4. 防火分区层压玻璃
5. 半透明玻璃
6. 外露混凝土
7. 背刚支撑的室外结构玻璃
8. 开启式窗户
9. 透明玻璃

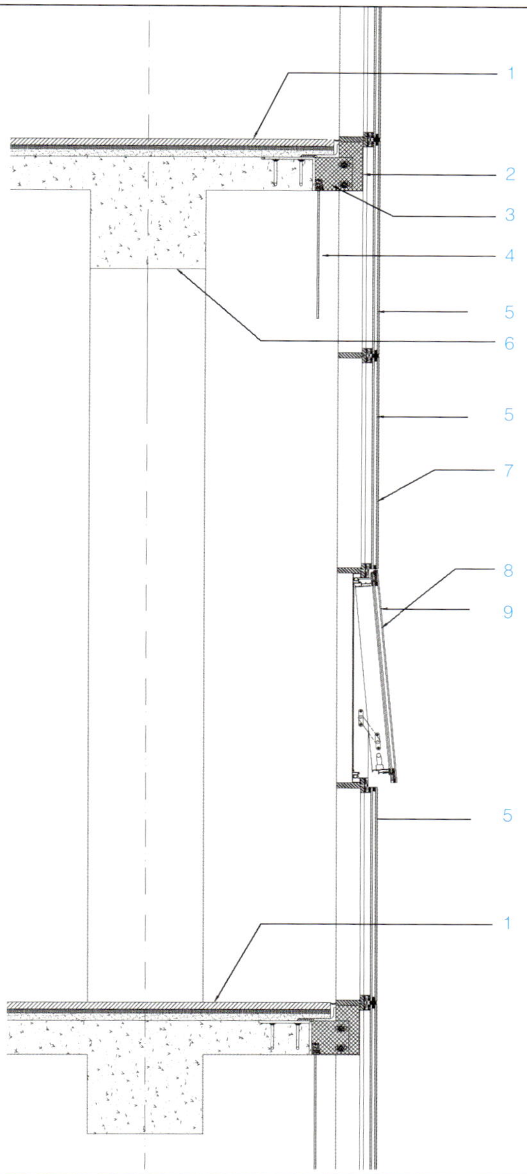

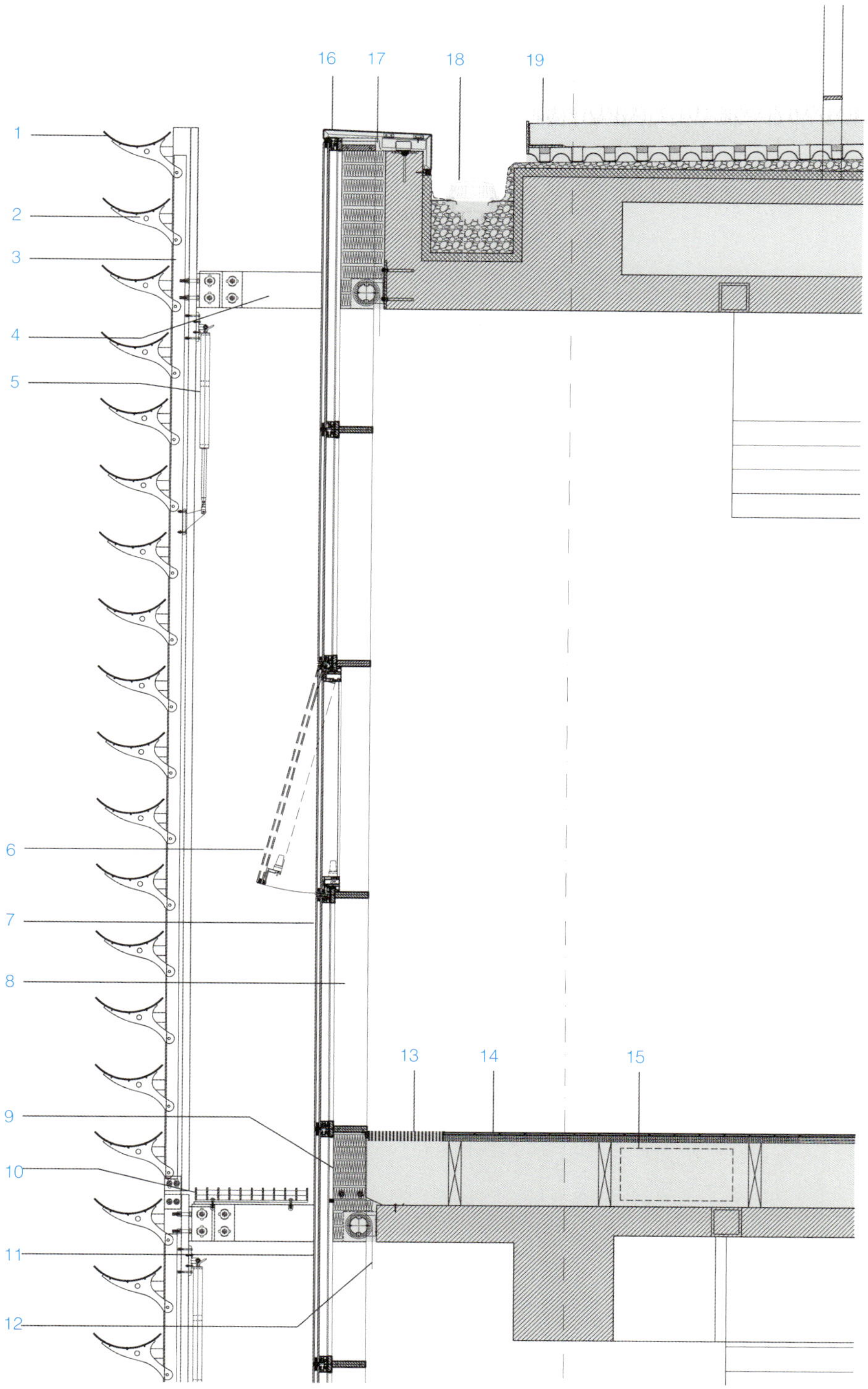

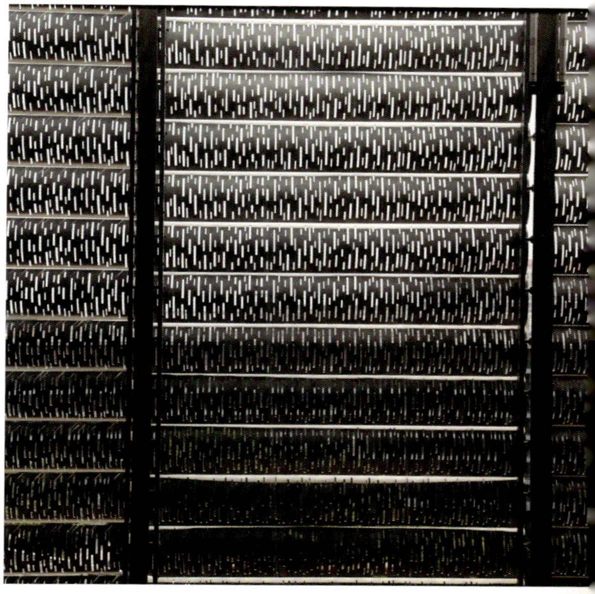

1. Operable perforated anodised aluminium louvres 275mmx2000mm
2. Annodised aluminium louvre bracket
3. Painted steel louvre frame
4. Painted steel bracket
5. Sensor controlled hydraulic piston
6. Operable window
7. Insulating glazing unit with solar control coating
8. Curtain wall frame
9. Aluminium cover with fire insulation infill
10. Painted steel catwalk
11. Insulationg glazing unit with acid etched fire resistant glass
12. Room darkening shade
13. Floor diffuser
14. Modular raised flooring system
15. Façade automation control box monitors interior/exterior environment sensor
16. Aluminium coping
17. Waterproofing menbrane
18. Drain
19. Modular planting system

1. 可操作阳极氧化铝穿孔百叶窗275毫米x2000毫米
2. 阳极氧化铝百叶窗支架
3. 有色钢天窗架
4. 有色钢支架
5. 传感器控制的液压活塞
6. 可操作窗口
7. 太阳能控制的绝缘涂层部件
8. 幕墙框架
9. 铝盖与防火隔热填充
10. 有色钢狭小通道
11. 绝缘玻璃系统中安置了酸蚀刻防火玻璃
12. 房间灯火管制
13. 楼层疏散区
14. 模块化活动地板系统
15. 立面自动化控制箱监控室内/室外环境传感器
16. 铝盖顶
17. 防水膜
18. 排水
19. 模块化种植系统

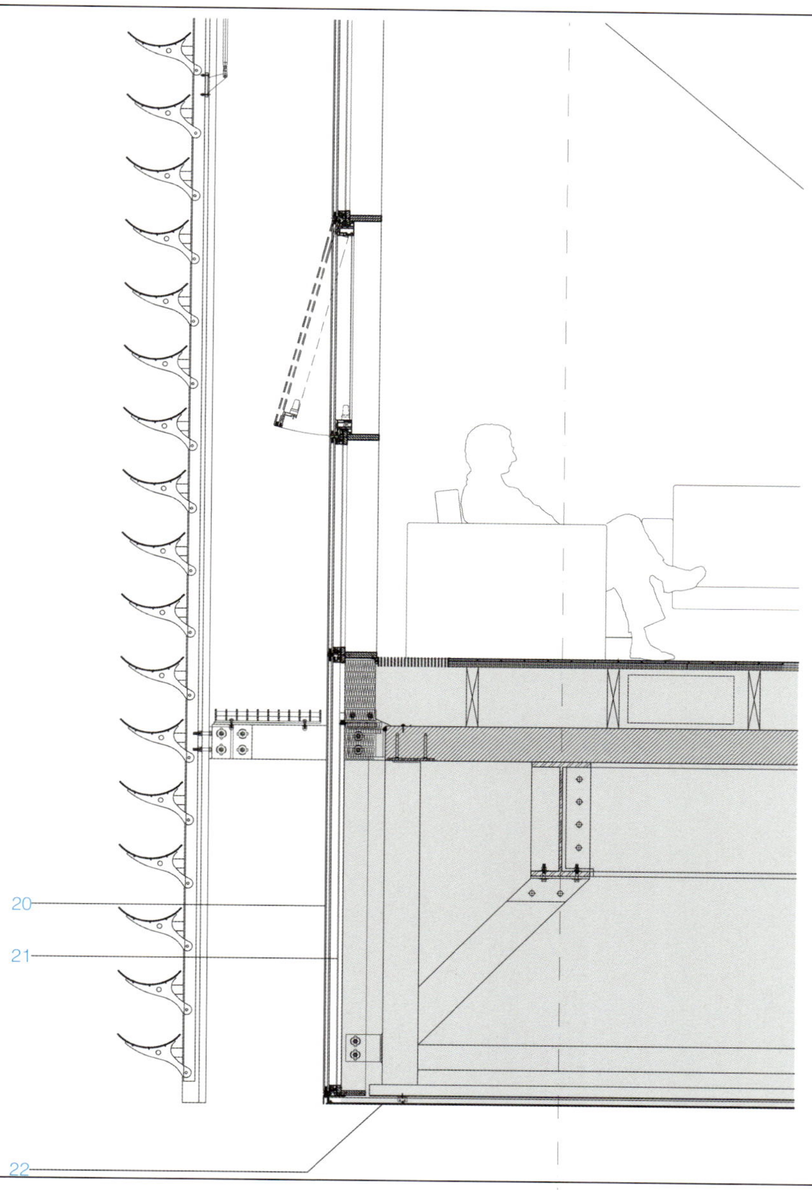

20. Insulating glazing unit
21. Metal panel shadow box
22. Painted aluminium soffit panels

20. 中空玻璃单元
21. 金属板玻璃框罩
22. 有色铝拱腹板

Swedbank Head Office Building

瑞典银行总部大楼

Location: Vilnius, Lithuania
Designers: Audrius Ambrasas Architects
Design team: Audrius Ambrasas, Vilma Adomonyte, Tomas Eidukeviius, Donatas Malinauskas
Photographers: R. Urbakavicius, A. Ambrasas, K. Satunas
Completion date: 2009

项目地点：立陶宛，维尔纽斯
设计师： Audrius Ambrasas建筑师事务所
设计团队：奥德琉斯・埃姆布拉瑟斯，维尔马・安东蒙耶特，托马斯・爱迪克维塞斯，多纳塔斯・马林纳斯卡斯
摄影师：R.俄贝卡维萨斯，A.埃姆布拉瑟斯，K.塞特纳斯
竣工时间：2009年

The Swedbank's new head office building in Vilnius is located in a fairly recently developed but rapidly growing political, commercial and recreational centre of the Lithuanian capital, which was moved from its historical centre to the right bank of the Neris River. The exclusive location of the building created a number of challenges. "We wanted the building to become an integral part of the development of the right bank of Neris River, completing it in an original way designing the space around it without becoming the centre of attention", Audrius Ambrasas said.
A distinctive feature of the new head office building is its openness and accessibility by the public. The site for the Swedbank's building was developed on the old Ukmerges Street which becomes the main axis of the buildings composition. The internal pedestrian street (the old Ukmerges Street) and the flowing spaces on the building's ground floor were planned as an public urban space. The building consists of two parts: high-rise part of two 15 and 16 storey high structures and the lower part comprises two more blocks: the broken-lined parterre and the regular four-storey administrative part. The highlight of the building is the over 4,500 metres terrace, constructed on the stylobate part and offering excellent views of the river bank.
The ground floor, under the terrace, opens a wide one-piece space. This artfully crafted part of the complex seeks to become an important public attraction centre even during cold seasons of the year. It comprises a café for 150 visitors, both employees and guests of the bank, and also an auditorium for an audience of 150 people and a reading room, in addition to the customer servicing centre and the business centre.
The developers and the designers demonstrate their respect to the public by using high-quality, long lasting and aesthetic materials and details. The most important - finishing of the blank façade walls performed using stainless-metal plates polished in different directions. The ornamental effect was obtained by varying the plate brushing direction what makes the façade look multi-coloured.

瑞典银行新总部大楼坐落于立陶宛行政中心维尔纽斯市经济、政治较为发达的地区，该地区是维尔纽斯市的商业和休闲中心，如今，维尔纽斯市的城市发展中心已逐渐由历史中心向里斯河右岸转移。这一特殊的地理位置为建筑的设计带来了诸多挑战。设计师奥德琉斯·埃姆布拉瑟斯曾这样说到："我们希望新总部大楼能够成为里斯河右岸总体规划中一个不可分割的部分，独特的外观能够与周边环境自然融合为一个完整的整体。"
开放和亲切是该总部大楼的一个鲜明特色。建筑所在的古老的Ukmerges大街成为整个建筑物的主轴。室内步行街（古老的Ukmerges大街）与一楼的流通空间作为城市公共空间。该建筑由两部分组成：其一是高层部分，涉及一个15层和一个16层的高层建筑；另一个则是中低层部分，涉及两个以上的建筑楼群，其中包括错落有致的花坛和较规则的四层行政大楼。占地4,500平方米的露台是整个建筑的亮点所在，面向柱基而设，将优美的河岸景致巧妙地引入到室内。
露台下方的一楼，向公众展开了一个开阔的一体式空间。这个精致的建筑区域成为一个重要的公众聚会中心，即使在寒冷的季节中此处也同样暖意融融。该中心中设置了一个同时容纳150人的自助餐厅，面向银行员工和办理业务的来宾而设；此外，还设有一个能够容纳150人的礼堂，以及一个阅览室和客户服务中心及商务中心。
开发商和设计师向人们展示了他们对高品质、耐久、美观材料及细节的注重。值得一提的是，单一的外立面墙体采用了抛光不锈钢金属板材料进行全方位设计。变换的金属板涂刷方向巧妙地营造出观赏效果，令建筑的立面五彩缤纷。

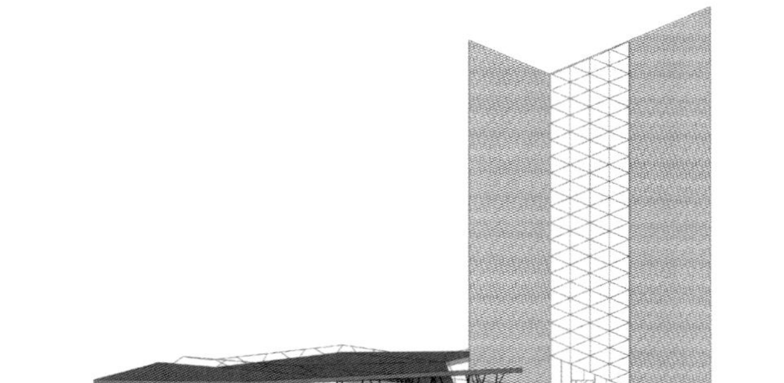

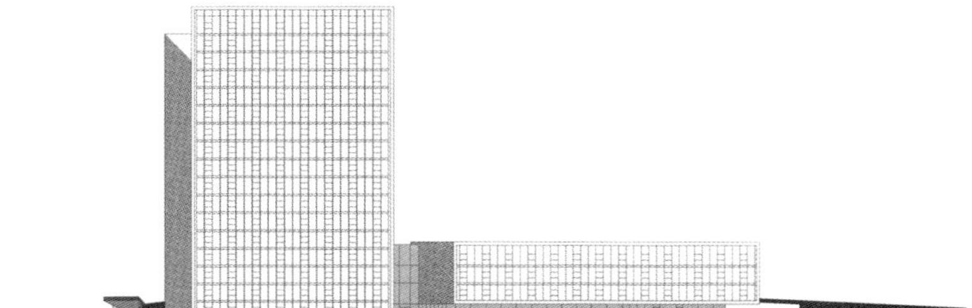

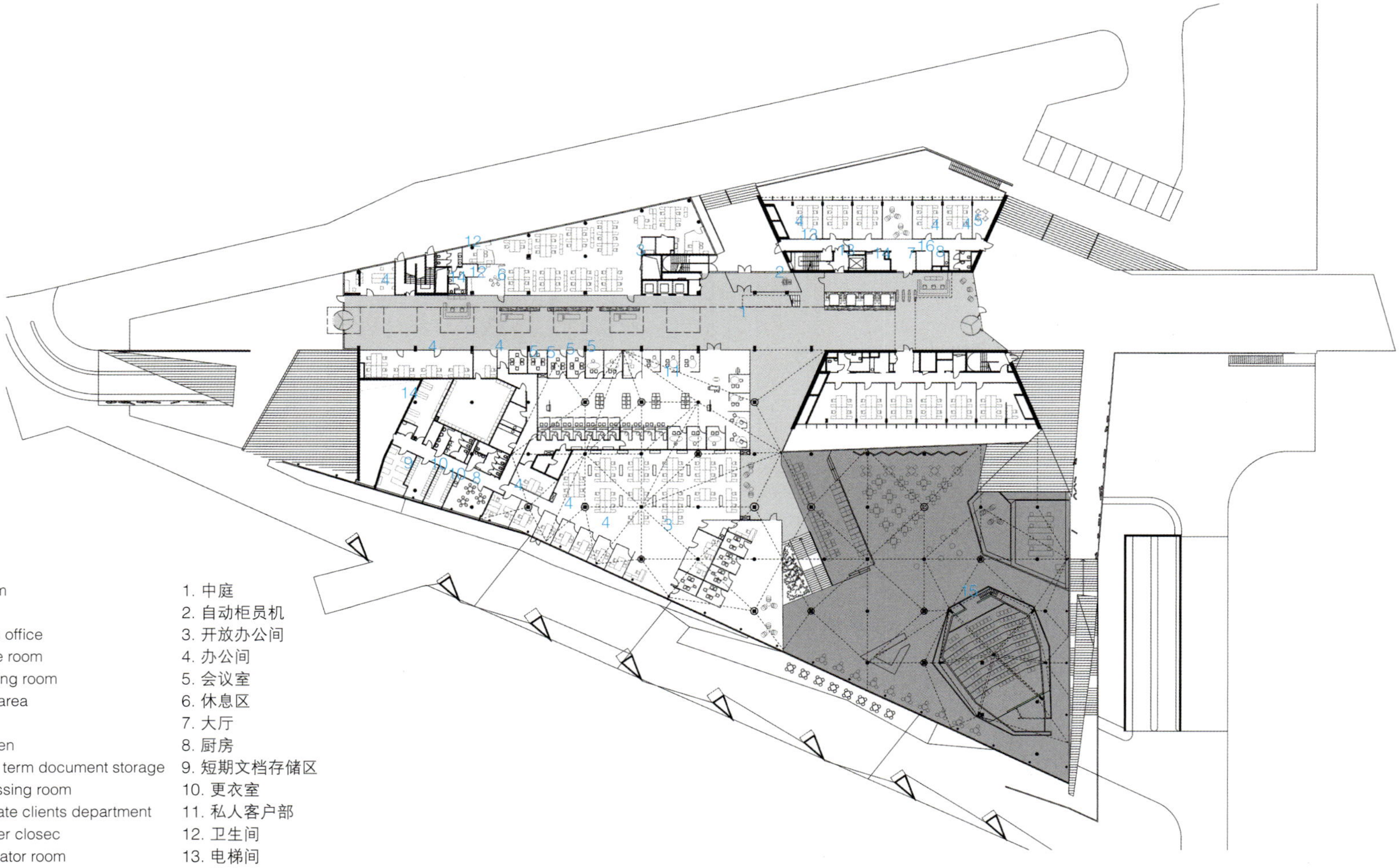

1. Atrium
2. ATM
3. Open office
4. Office room
5. Meeting room
6. Rest area
7. Hall
8. Kitchen
9. Short term document storage
10. Dressing room
11. Private clients department
12. Water closec
13. Elavator room
14. Technical room
15. Balcony
16. Copy room

1. 中庭
2. 自动柜员机
3. 开放办公间
4. 办公间
5. 会议室
6. 休息区
7. 大厅
8. 厨房
9. 短期文档存储区
10. 更衣室
11. 私人客户部
12. 卫生间
13. 电梯间
14. 技术室
15. 阳台
16. 复印室

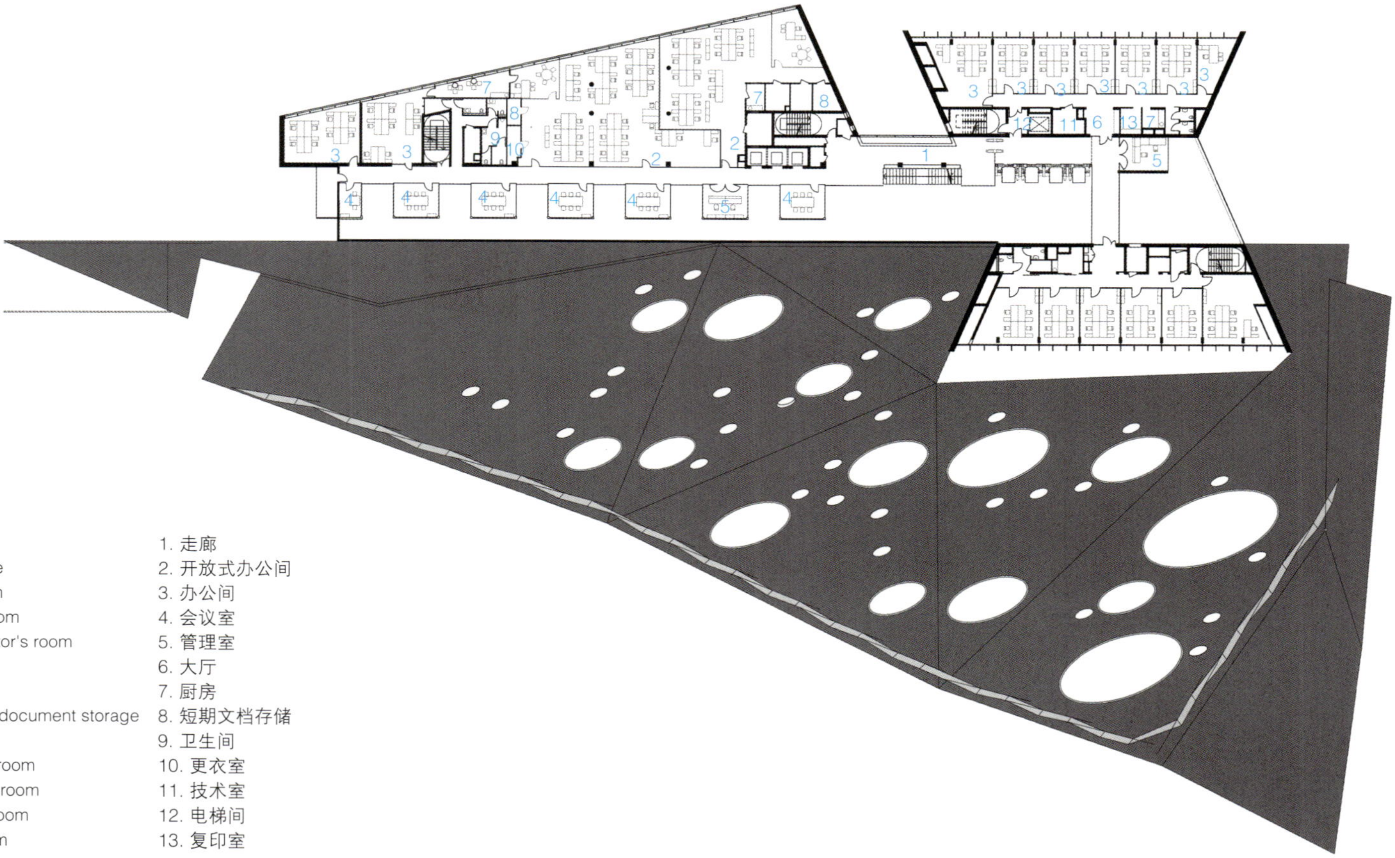

1. Gallery
2. Open office
3. Office room
4. Meeting room
5. Administrator's room
6. Hall
7. Kitchen
8. Short term document storage
9. WC
10. Dressing room
11. Technical room
12. Elavator room
13. Copy room

1. 走廊
2. 开放式办公间
3. 办公间
4. 会议室
5. 管理室
6. 大厅
7. 厨房
8. 短期文档存储
9. 卫生间
10. 更衣室
11. 技术室
12. 电梯间
13. 复印室

Swedbank

Swedbank

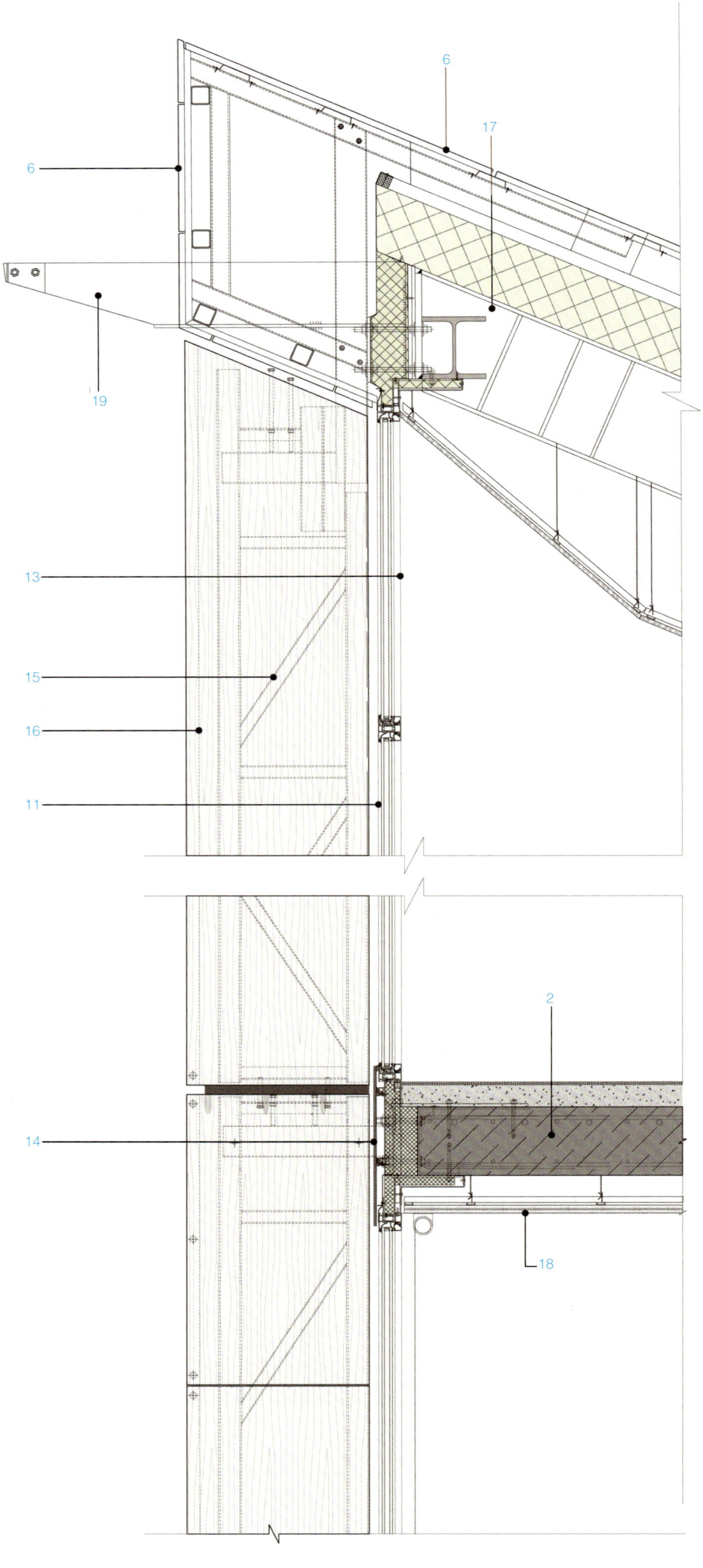

1. Reinforced concrete wall
2. Reinforced concrete slab
3. Steel hinge
4. Mounting angle (galvanised)
5. Aluminium plate
6. Decorative stainless steel (0.8mm)
7. Thermal insulation 150mm
8. Thermal insulation 30mm
9. Metal frame structure
10. Aluminium profiles
11. Glass Ipasol Sky 50/26, 8mm
12. Clear float glass, 8mm
13. Transparent sound-insulating laminated glass, 8mm
14. Shatterproof glass "Blackpearl", 8mm
15. Profiles (galvanised frame)
16. Plate Prodema Baq +, 8mm
17. Steel structure of the roof
18. Ceiling - double gypsum board on metal shell
19. Stainless steel bracket for cleaning equipment

1. 钢筋混凝土墙
2. 钢筋混凝土板
3. 钢铰链
4. 安装角度（镀锌）
5. 铝板
6. 装饰不锈钢（0.8毫米）
7. 150毫米隔热墙
8. 30毫米隔热墙
9. 金属框架结构
10. 铝型材
11. 玻璃"Ipasol Sky" 50/26型材，8毫米
12. 透明浮法玻璃，8毫米
13. 透明隔声夹层玻璃，8毫米
14. "黑珍珠"不碎玻璃，8毫米
15. 剖面图（镀锌框）
16. 8毫米厚外墙薄膜板
17. 屋顶钢架结构
18. 金属外壳上的双重吊顶石膏板
19. 清洗设备的不锈钢支架

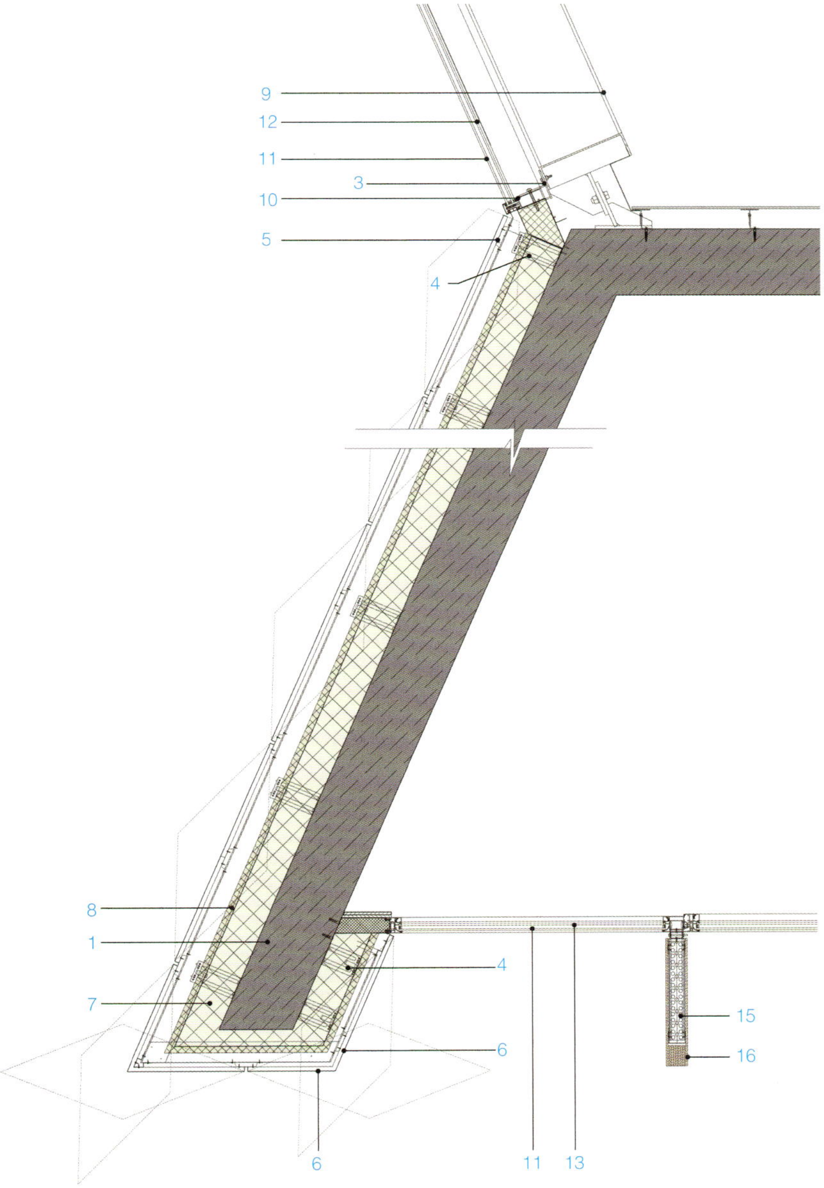
9
12
11
3
10
5
4
8
1
4
7
6
6
11
13
15
16

Swedbank

BDP Manchester Studio

BDP曼彻斯特工作室

Location: Manchester, UK
Designers: BDP
Owner: Town Centre Securities
Completion date: 2008

项目地点：英国，曼彻斯特
设计师：BDP建筑师事务所
业主：Town Centre Securities公司
竣工时间：2008年

Award name:
MIPIM Award (Green Building - Finalist) 2009
RIBA Award 2009
IStructE Regional Award (Best Sustainable Project) 2009
RICS Regional Award (Highly commended sustainability category) 2009
British Council for Ofces Award (Regional Award) 2009
Manchester Society of Architects Award 2009
Roses Design Award (Bronze) 2009

获奖名称：
2009年国际房地产投资展览会奖（绿色建筑－决赛入围者）
2009年英国皇家建筑师学会奖
2009年国际组织结构工程师协会区域奖（最佳可持续发展项目）
2009年英国皇家特许测量员协会区域奖（可持续性类项目二等奖）
2009年英国文化协会办公室奖（区域奖）
2009年曼彻斯特建筑师学会奖
2009年玫瑰设计奖（铜奖）

The six-storey building on Ducie Street was developed by Town Centre Securities as part of their Piccadilly Basin masterplan. Situated along the southern edge of the canal basin, the building is used exclusively by BDP's Manchester team.
The building provides large open plan studio space and ancillary accommodation including a hub space at ground floor level. This interactive area including café, staff restaurant and extended reception space, overlooks the canal at raised ground level.
A striking feature of the building is the punctuated stainless-steel south façade that rises above the Ducie Street colonnade to contain the open-plan studio areas before sweeping over to form the roof of the building. The reflective external finish, heavily insulated build-up and narrow vertical apertures all serve to minimise solar heat gain, and to maximise privacy with the residential buildings opposite.
By contrast, the northern façade of the building is transparent. The floor-to-soffit glazing takes maximum advantage of north light to illuminate the full extent of studio spaces and reveals wonderful views of the city centre. A fully-glazed circulation staircase cantilevered over the canal provides the circulation for all floors.
Sustainability has been a key driver in all aspects of the design and delivery of the new studio which is an expressive response to context and microclimate. Rainwater is harvested from the roof and used to flush toilets throughout the building. It is the first naturally ventilated and night time cooled office building in Manchester to achieve an Excellent BREEAM rating.

位于迪西大街上的BDP曼彻斯特工作室共六层，是皮卡迪利湾总体规划的组成部分。该建筑由Town Centre Securities公司兴建，仅供BDP曼彻斯特团队使用。
建筑内设置了开阔的工作室和附属设施空间，其中附属设施空间中包括一个一层的中心区。这一互动区中设有咖啡厅、员工餐厅和俯瞰运河的接待中心。
位于迪西大街一侧的南部不锈钢金属立面是整个建筑的亮点所在，开放式的工作室空间与屋顶形成完美的弧线型外观。外部的抛光反射材料具有良好的绝缘和隔热效果，狭窄的垂直开口能够有效地减少室内太阳热量的获取，并为建筑后身的住宅空间提供完美的私密环境。
与南端立面不同，建筑的北侧立面较为通透。无论是地面还是天花板均充分利用了来自北部的自然光线，最大程度地确保工作室空间光线充足、通透，并能够将城市中心的美景在眼底尽收。楼梯口采用全玻璃设计，悬垂于运河之上，为各楼层间的流通提供条件。
可持续性是整个设计的关键，以应对建筑周围的环境和微气候。屋顶上蓄积的雨水可用来冲洗厕所。该建筑是曼彻斯特首个采用自然通风、夜间制冷的办公建筑，也因此获得了绿色建筑评价标准优秀等级。

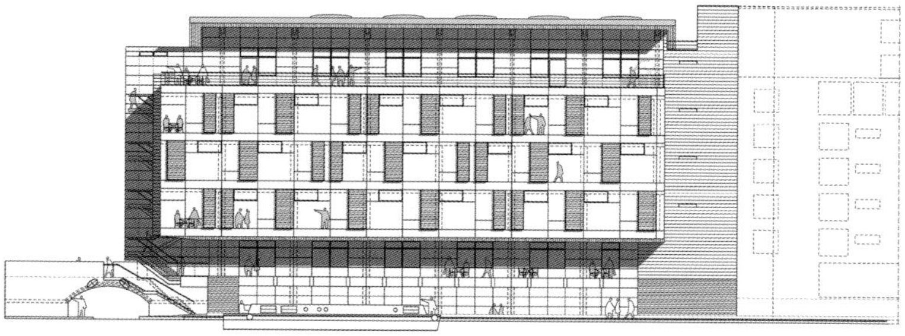

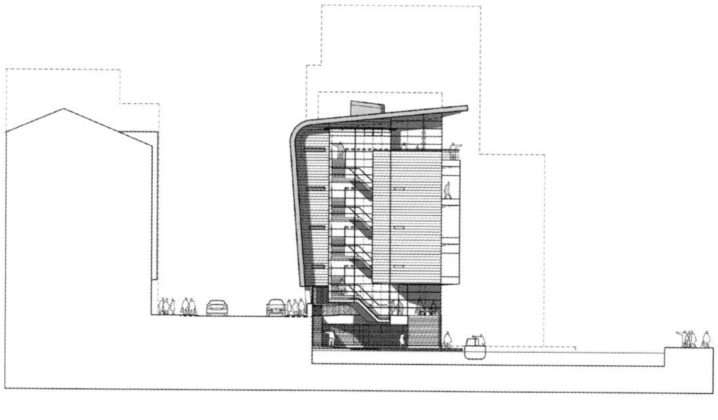

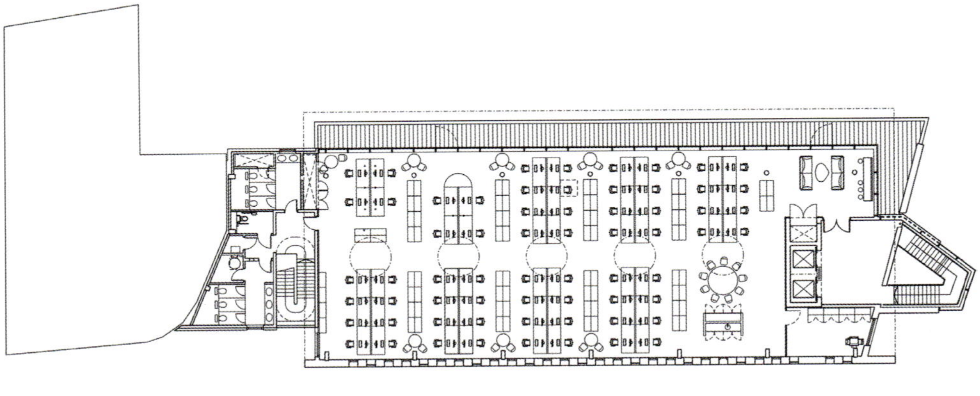

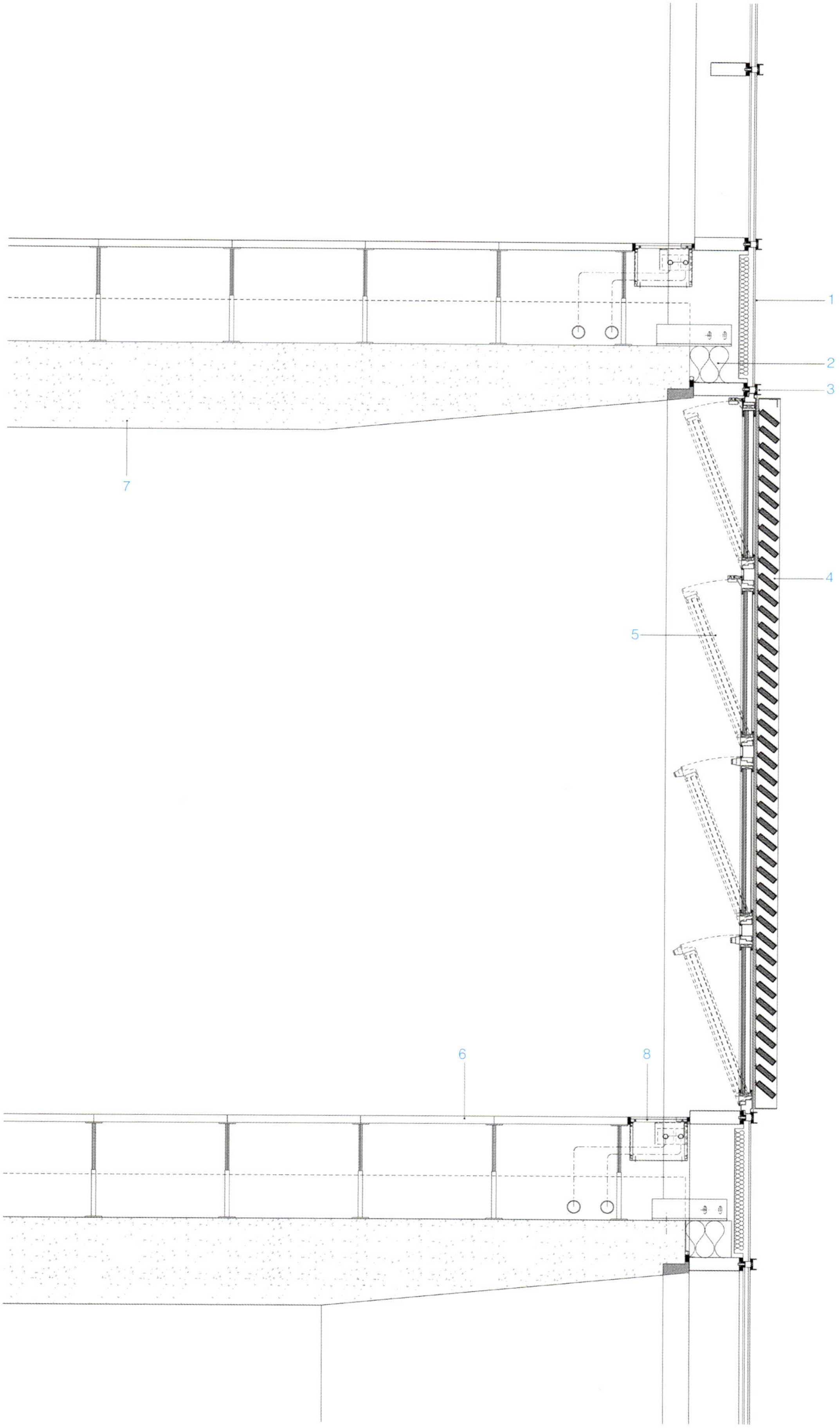

1. Opaque infill panel-100% fritted glass with integral insulated backing concealing raised floor zone&concrete floor edge
2. 90-minute mineral wool fire stop cavity barrier
3. Schuco FW60 curtainwall system in Syntha Pulvin PPC ral 9006 internal frame with anodised aluminium 'C' profile cover cap and through face star head fixing
4. Louvred cassette comprising 90mmx18mm cedar blades supported by bespoke PPC aluminium frame, fixed into curtainwall system
5. Hopper vents comprising Schuco window units with aluminium faced infill and mineral wool insulation. Inner visible face, ribbed profile PPC aluminium from Goodings Aluminium. Two uppermost vents fitted with actuators by Dyer, lower vents manual with Schuco gearbox handles
6. Raised access floor
7. Post tensioned concrete floor slab
8. Floor heater box/grille

1. 不透明填充板－100%多孔玻璃，设有整体隔热衬垫，将活动底板和混凝土地板的边缘地带进行隐藏
2. 90分钟耐火消防屏障
3. 旭格FW 60幕墙系统采用Syntha Pulvin PPC ral 9006作为内部框架，阳极氧化铝罩盖，通过“星头”进行固定
4. 固定到幕墙系统上的百叶窗匣中由90x18毫米的雪松百叶和铝合金框架构成
5. 倒开气窗由Schuco窗口、铝合金窗框以及矿物棉绝缘层构成。可见内面、肋剖面以古丁斯铝为材料。两个最高的通风口安装了戴尔驱动器，低处的通风口则采用了旭格变速箱进行手动控制
6. 活动地板
7. 后张预应力混凝土楼板
8. 地板加热箱/格栅

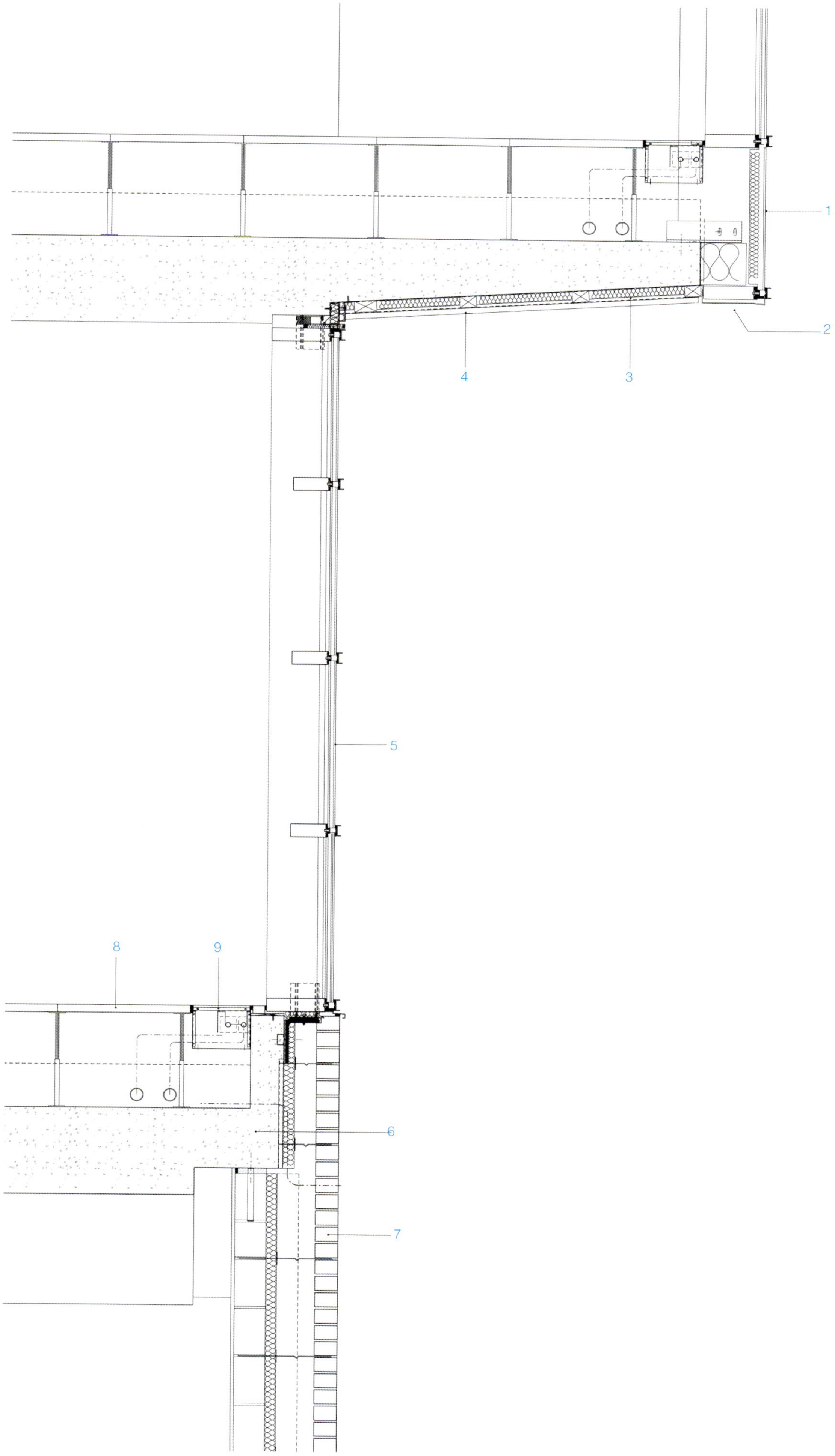

1. Opaque infill panel-100% fritted glass with integral insulated backing concealing raised floor zone & concrete floor edge
2. Glased in anodised aluminium profile cloak wrapping around soffit of mullion
3. Vapour Barrier, soffit insulation and breather menbrane
4. Cedar soffit boarding pinned to 100x50 & 75x50 plugged &screwed to concrete soffit
5. Schuco FW60 curtainwall system in Syntha Pulvin PPC ral 9006 internal frame with anodised aluminium 'C' profile cover cap and through face star head fixing
6. Post tensioned concrete floor slab
7. Newland blend facing brick with flush pointed mortor
8. Raised access floor
9. Floor heater box/grille

1. 不透明填充板 – 100%多孔玻璃，设有整体隔热衬垫，将活动底板和混凝土地板的边缘地带进行隐藏
2. 阳极氧化铝型材环绕窗子竖框的拱腹
3. 蒸汽屏蔽，底部绝缘和呼吸薄膜
4. 雪松拱腹板用100x50和75x50的栓塞固定到混凝土拱腹之上
5. 旭格FW 60幕墙系统采用Syntha Pulvin PPC ral 9006作为内部框架，阳极氧化铝罩盖，通过“星头”进行固定
6. 后张预应力混凝土楼板
7. 带有闪点的新大陆混合贴面砖
8. 活动地板
9. 地板加热箱/格栅

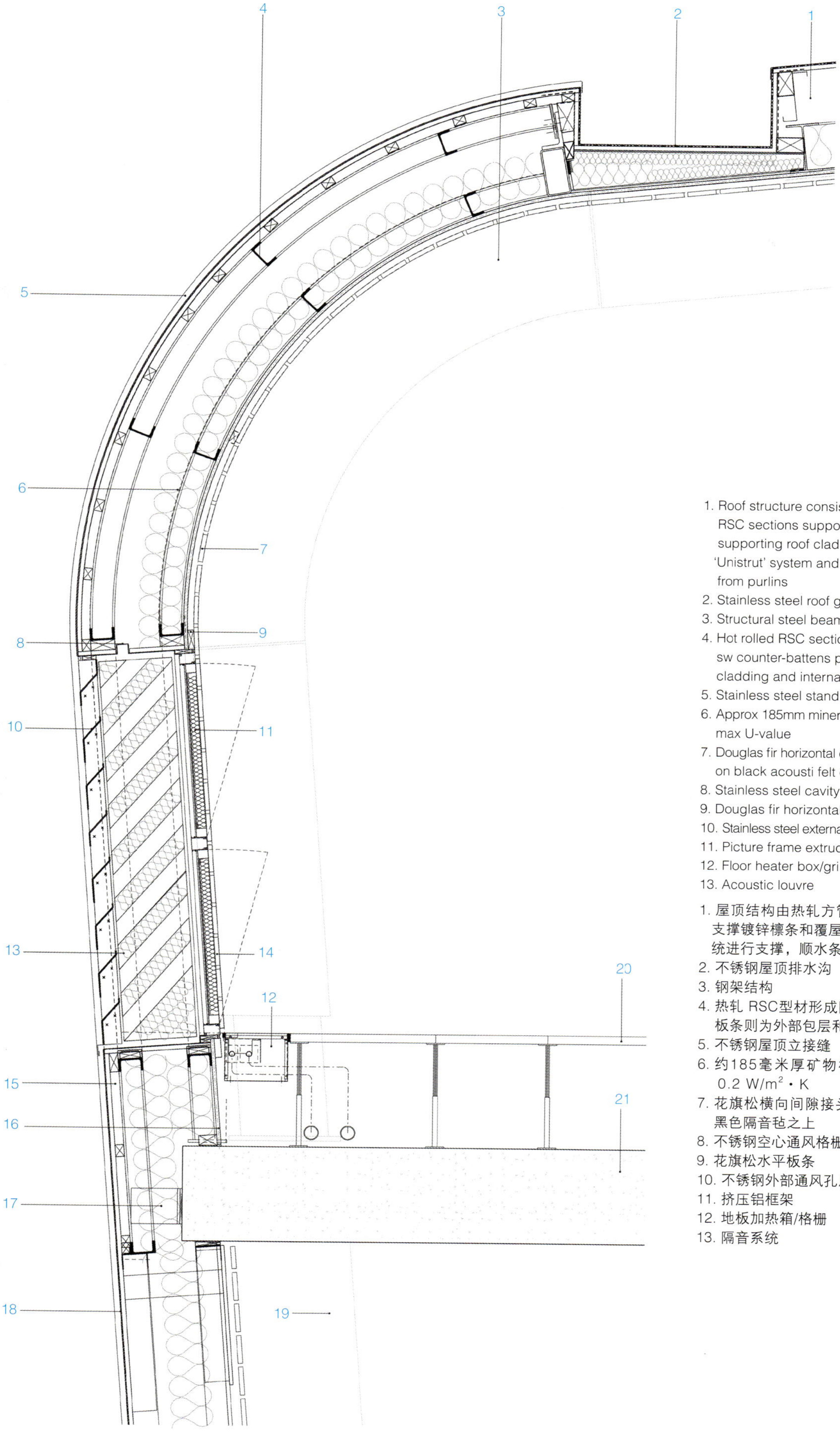

1. Roof structure consisting of hot rolled RHS and RSC sections supporting galvanised 'Z' purlins supporting roof cladding. Ceiling supported on 'Unistrut' system and counter battens suspended from purlins
2. Stainless steel roof gutter
3. Structural steel beam
4. Hot rolled RSC sections forming radiused eaves, sw counter-battens providing fixing for external cladding and internal lining
5. Stainless steel standing seam roof
6. Approx 185mm mineral wool insulation 0.2 $W/m^2 \cdot K$ max U-value
7. Douglas fir horizontal open jointed timber boarding on black acousti felt on timber battens
8. Stainless steel cavity ventilation grille
9. Douglas fir horizontal batten
10. Stainless steel external louvre with bird mesh to rear
11. Picture frame extruded aluminium
12. Floor heater box/grille
13. Acoustic louvre
14. Natural ventilation unit comprising of two bottom hung double gasket air tight insulated hopper door units. Upper hopper controlled by actuator, lower hopper manual opening. Douglas fir frames &douglas fir veneered plywood hopper door panels
15. 18mm thick external quality ply fix to 38x50 horizontal treated softwood timber rails fixed to channel studs
16. 18mm ply panel below raised floor on acoustic sheet membrane on 15mm thick plasterboard lining on timber battens. Silicone sealant between internal lining and structure
17. 90-minute fire stop cavity barrier
18. Exterior quality stainless steel cladding over breather membrane on exterior grade wbp plywood sheeting
19. Concrete column
20. Raised access floor
21. Concrete floor slab

1. 屋顶结构由热轧方管和防翻滚稳定系统型材支撑镀锌檩条和覆屋面构成。天花板由焊件系统进行支撑，顺水条从檩条上悬垂下来
2. 不锈钢屋顶排水沟
3. 钢架结构
4. 热轧 RSC型材形成圆角屋檐，而相反方向的板条则为外部包层和内部衬垫提供固定区域
5. 不锈钢屋顶立接缝
6. 约185毫米厚矿物棉绝缘层，导热系数为 0.2 $W/m^2 \cdot K$
7. 花旗松横向间隙接头木板位于木板条上方的黑色隔音毡之上
8. 不锈钢空心通风格栅
9. 花旗松水平板条
10. 不锈钢外部通风孔后方设有防鸟网
11. 挤压铝框架
12. 地板加热箱/格栅
13. 隔音系统
14. 自然通风装置由两个底部悬垂双重垫片密封绝缘底开门构成，上方的底开门由制动器进行控制，下方的则采用手动开启。花旗松框架，底开门板材为花旗松贴面胶合板
15. 18毫米厚的外层胶合板固定于38x50水平处理软木肋骨之上，而软木肋骨则与凹槽螺柱固定在一起
16. 活动地板下方的胶合板隔音薄膜之上，15毫米厚的石膏板位于隔音薄膜的下方、木板条的上方。内部衬里和结构之间由硅酮密封剂进行密封
17. 90分钟耐火屏障
18. 外部酚醛胶合板上的通气薄膜覆以不锈钢表层
19. 混凝土柱
20. 活动地板
21. 混凝土楼板

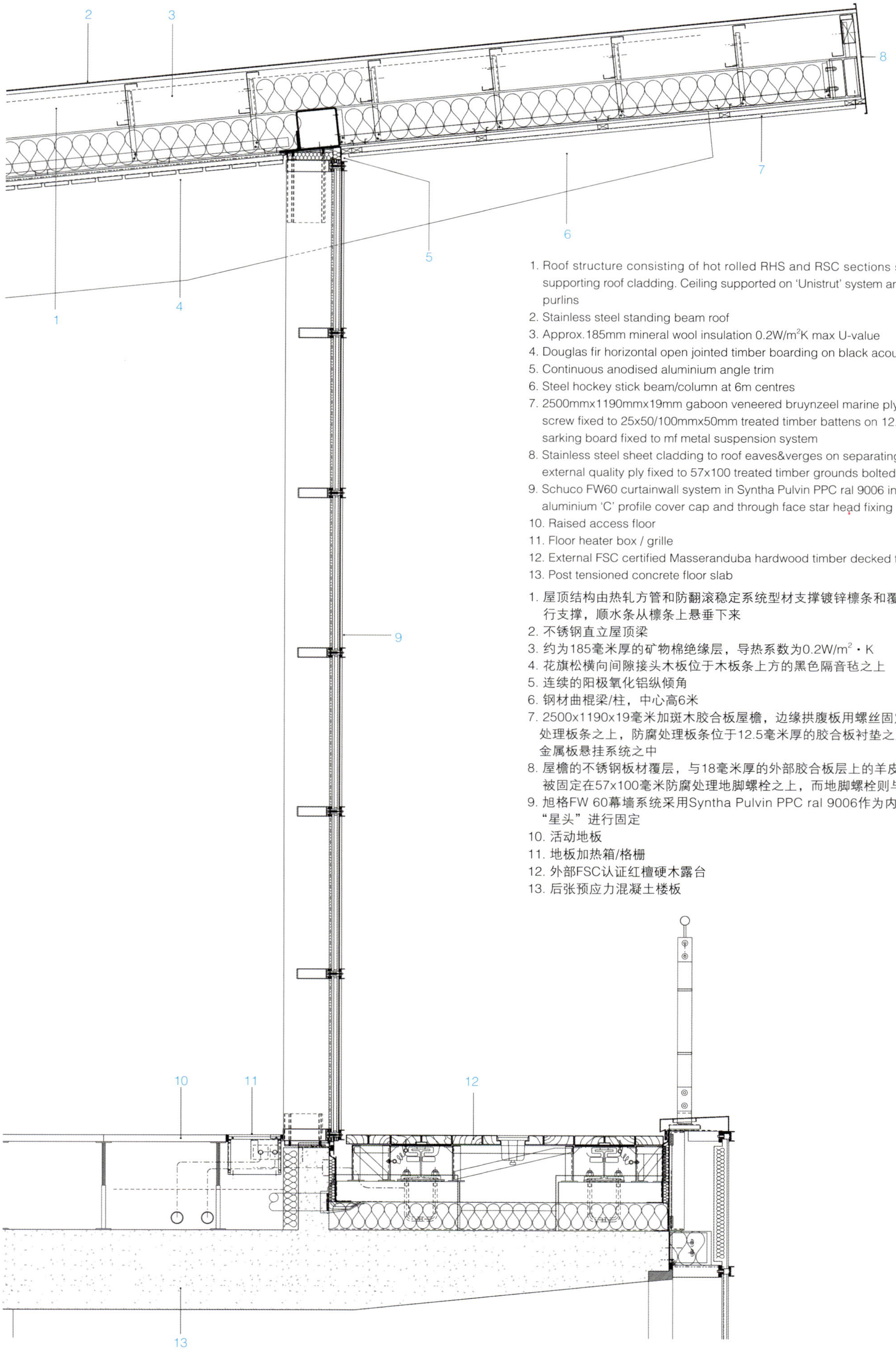

1. Roof structure consisting of hot rolled RHS and RSC sections supporting galvanized 'Z' purlins supporting roof cladding. Ceiling supported on 'Unistrut' system and counter battens suspended from purlins
2. Stainless steel standing beam roof
3. Approx.185mm mineral wool insulation 0.2W/m^2K max U-value
4. Douglas fir horizontal open jointed timber boarding on black acoustic felt on timber battens
5. Continuous anodised aluminium angle trim
6. Steel hockey stick beam/column at 6m centres
7. 2500mmx1190mmx19mm gaboon veneered bruynzeel marine ply roof eaves / verge soffit panels screw fixed to 25x50/100mmx50mm treated timber battens on 12.5mm thick external quality plywood sarking board fixed to mf metal suspension system
8. Stainless steel sheet cladding to roof eaves&verges on separating fleece membrane on 18mm thick external quality ply fixed to 57x100 treated timber grounds bolted to purlins at 450mm centres
9. Schuco FW60 curtainwall system in Syntha Pulvin PPC ral 9006 internal frame with anodised aluminium 'C' profile cover cap and through face star head fixing
10. Raised access floor
11. Floor heater box / grille
12. External FSC certified Masseranduba hardwood timber decked terrace
13. Post tensioned concrete floor slab

1. 屋顶结构由热轧方管和防翻滚稳定系统型材支撑镀锌檩条和覆屋面构成。天花板由焊件系统进行支撑，顺水条从檩条上悬垂下来
2. 不锈钢直立屋顶梁
3. 约为185毫米厚的矿物棉绝缘层，导热系数为0.2W/m^2·K
4. 花旗松横向间隙接头木板位于木板条上方的黑色隔音毡之上
5. 连续的阳极氧化铝纵倾角
6. 钢材曲棍梁/柱，中心高6米
7. 2500x1190x19毫米加斑木胶合板屋檐，边缘拱腹板用螺丝固定在25x50/100x50毫米厚的防腐处理板条之上，防腐处理板条位于12.5毫米厚的胶合板衬垫之上，胶合板衬垫用螺丝固定于MF金属板悬挂系统之中
8. 屋檐的不锈钢板材覆层，与18毫米厚的外部胶合板层上的羊皮纸薄膜分离开来，外部胶合板层被固定在57x100毫米防腐处理地脚螺栓之上，而地脚螺栓则与中心高为450毫米的檩条相连
9. 旭格FW 60幕墙系统采用Syntha Pulvin PPC ral 9006作为内部框架，阳极氧化铝罩盖，通过“星头”进行固定
10. 活动地板
11. 地板加热箱/格栅
12. 外部FSC认证红檀硬木露台
13. 后张预应力混凝土楼板

Cabel Industry

Cabel Industry 软件公司总部

Location: Empoli, Italy
Designer: Massimo Mariani
Design team: Elda Bellone / Roseda Gentile / Alessandro Mariani / Giovanni Lunari / Simona Baronti
Photographer: Alessandro Ciampi
Completion date: 2008

项目地点：意大利，恩波利
设计师：马西莫·马里亚尼
设计团队：埃尔达·贝洛内，罗丝达·詹蒂莱，亚历山德罗·马里亚尼，乔瓦尼·卢纳里，西蒙娜·巴龙蒂
摄影师：亚历山德罗·勃朗迪
竣工时间：2008年

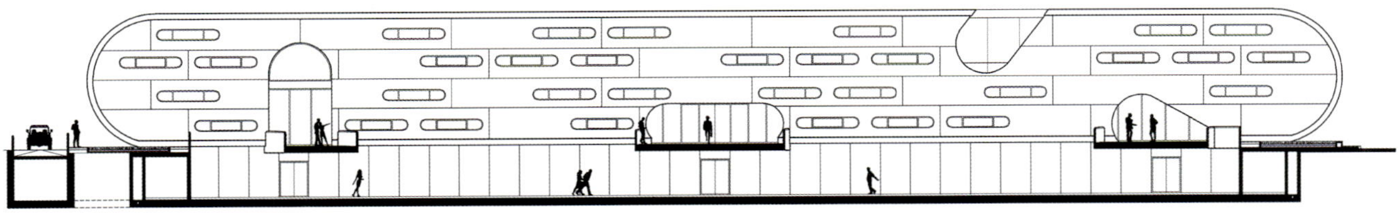

Just out the edge of the town of Empoli, the building is the headquarters of Cabel Industry (a company dealing in computer systems for banks), it covers an area of approximate 4,500 square metres over several levels and it is incorporated on the small local industrial estate setting up new dialectical relations with the local manufacturing fabric.
Commitment asked the designers to project an office building to be constructed using industrial methods by keeping down costs and time of construction. They decided to design a new precast concrete panel featured to be stand on the two main façades. Partially set into the ground, the building is composed of two extended floors out the ground level and a vault under. Along the main front the façade is protected by a long strip of public landscaping running parallel to the road.
The visitor arrives on the ground level through three suspended bridges launched on a large excavation which lights the vault designed to house expositions and art installations. At night time this empty space becomes a pool of light which allows the architecture to detach itself from the ground, making it look like a lightweight multi-coloured object suspended in the dark.
Inside the building the underground level holds a printing facility and other spaces allocated for various different kinds of activities; the ground floor features all different sorts of glazed spaces, ranging from open plan spaces to small isolated cells, in a modern interaction of liquid working areas like acquarium. In contrast the first floor accomodates the management space, with a small internal patio and terrace in-between them. The elevators and distributional stairs take up the middle section of the structural grid, made of stone tiles with steel elevators.
The gently curving building shell, windows, entrance gaps and furniture decoration all feature the same basic stylistic design.
During the day coloured-glass create liquid chromatic effects black and white offices inside, instead of night time when coloured cuts project out vivid lighting effects underlining holes, cuts and shapes of the building.
The building was constructed out of prefabricated concrete elements (painted with white protective enamel) and its roof and end sections were tiled in grey aluminium. The entire roof is covered with a system of photovoltaic solar panels using amorphous polycrystalline technology, carefully positioned so as not be visible. Thanks to this system, which is capable of generating approximately 150 kilowatts, the building is almost totally self-sufficient from an energy viewpoint.

位于恩波利市郊区的Cabel Industry 软件公司总部（该公司主要负责经销银行电脑系统软件），占地面积约4.500平方米，与小型工业园区的设计风格完美融合，并为该地区的工业建筑注入了无限活力。
客户希望该建筑能够通过工业设计手法尽可能地降低成本并在短时间内完成，因此，建筑主体采用了钢筋混凝土与铝制表层，地面两层，地下一层。外立面被一个平行于公路的狭长景观带所围绕。
地下一层的访客穿过三个设于一个大型艺术品展示中心之上的悬索桥即可到达地面层。夜幕降临之时，该空间即变成一个照明池使建筑与地面分离开来，遥看过去犹如一个轻盈的彩色发光体悬浮于夜色之中。
地下楼层中设有印刷设备，并提供多样化的活动空间；地面一层空间中则分布了各种玻璃空间，无论是开放空间还是独立小室，完美地衔接在一起，犹如一个美轮美奂的水族馆，极富现代气息。二层空间中设有管理空间，中间伴有室内天井和露台。电梯和楼梯位于中央，采用钢材和石砖材料。柔和的流线型建筑外观与窗口、入口以及家具装饰的外形设计相得益彰。
在白天，彩色玻璃为白色和黑色办公空间注入了绚丽色彩，当夜幕降临之时，从室内散发出的彩色光线完美地勾勒出外立面的孔洞、开口以及建筑外形。
建筑主体采用预制混凝土材料（外部覆以白色瓷釉起到保护作用），屋顶和端截面则采用了灰色铝制表层。整个屋顶经过精心部署，覆盖光伏太阳能电池板系统。这一设计，能够产生约150万千瓦能量，该建筑从能源的观点看几乎是完全实现了自给自足。

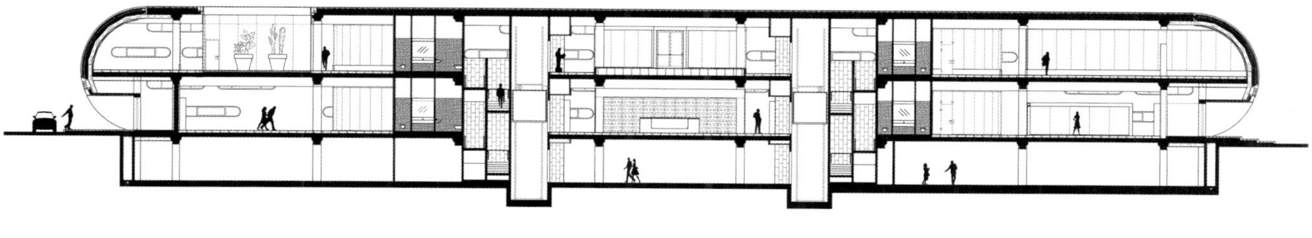

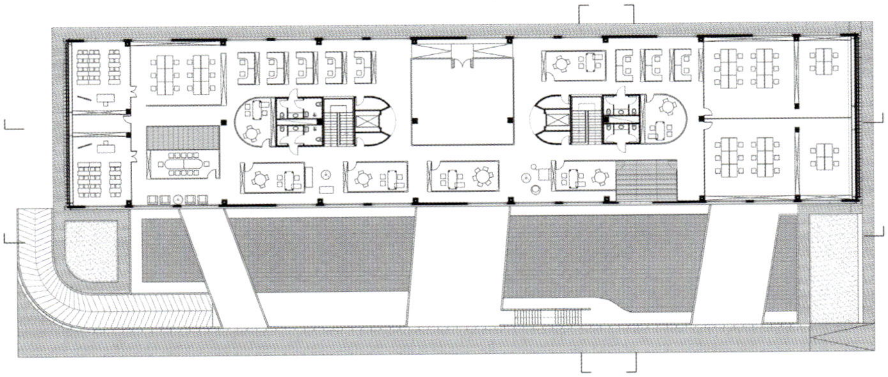

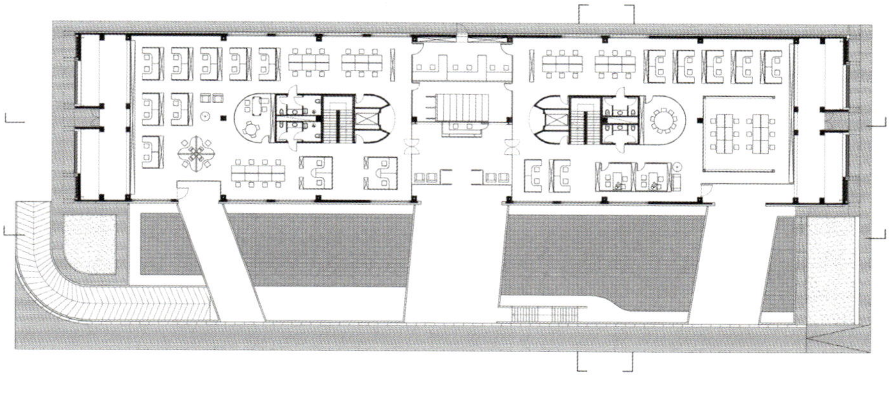

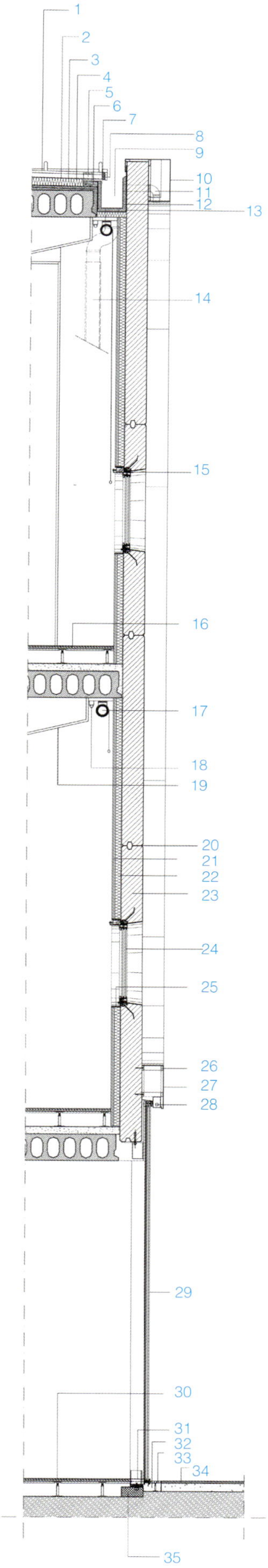

1. Fixing point by waterproof screw
2. Steel reinforcing 50/5 mm
3. Metallic hook system
4. Steel channel
5. Aluminium channel
6. Aluminium cladding frame
7. 108mmx55 mm steel post-and-rail structure
8. Aluminium fixing plate on glass structure post
9. Aluminium tubular section 50mmx50mm

1. 防水螺丝固定点
2. 型钢混凝土50 / 5 毫米
3. 金属挂钩系统
4. 槽钢
5. 铝材通道
6. 铝包层框架
7. 108x55毫米的钢柱及围栏结构
8. 玻璃结构标杆上的铝材固定板
9. 铝合金管剖面50x50毫米

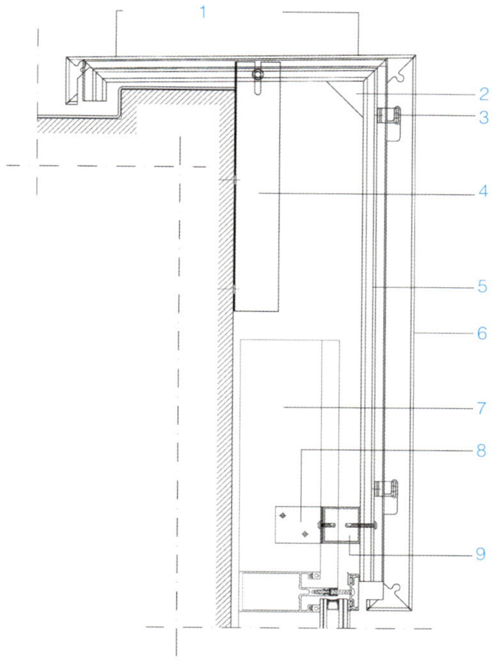

1. Aluminium roof
2. Mineral wool insulation thickness 100mm
3. Reinforced concrete layer pending 2.6%
4. Precast concrete floor
5. Fixing clip for roof cladding
6. Steel omega section
7. Galvanised steel angle
8. Polythene plug
9. Inox steel gutter
10. Aluminium cladding
11. Drainage tube
12. Insulation
13. External stainless steel parapet-gutter
14. Rainwater drainage tube inside the reinforced concrete pillar
15. Aluminium fastening type SHUCO vasistas opening
16. Floating floor made of marble conglomerate tiles, dimensions 600mmx600mm, thickness 30mm
17. Internal sun screen
18. Continuous light appliance
19. Plasterboard panel
20. Silicone
21. Plasterboard white painted
22. Thermal insulation 60mm
23. Pre-cast concrete panel
24. Aluminium fastening with with double glazing: laminate safety glass+cavity 16mm+laminate safety glass
25. 2mm steel cladding frame with paint finish
26. Steel structure for aluminium carter
27. 3mm sheet aluminium
28. Continuous lighting appliance SLOTLIGHT
29. Fixed aluminium fastening with double glazing: laminated retro-painted safety glass+cavity+laminated safety glass
30. Floating floor made of marble conglomerate tiles, dimensions 600mmx600mm, thickness 30mm
31. Aluminium tubular section with paint finish, 125mmx50mm
32. Stainless steel parapet-gutter
33. Stainless steel grill
34. External concrete pavement
35. Concrete threshold

1. 铝合金屋顶
2. 岩棉保温层厚度100毫米
3. 钢筋混凝土悬层2.6%
4. 预制混凝土楼板
5. 屋顶包覆固定扣
6. 欧米加高硅钢材
7. 镀锌角钢
8. 聚乙烯插头
9. 不锈钢排水沟
10. 铝覆层
11. 引流管
12. 绝缘
13. 外部不锈钢护栏，排水沟
14. 钢筋混凝土柱内的雨水引流管
15. 铝材紧固型SHUCO 气窗开口
16. 由大理石砾岩瓷砖制成的浮动地板，尺寸600x600毫米，厚度30毫米
17. 内部遮阳屏
18. 连续照明设备
19. 石膏板
20. 有机硅21白色石膏板
21. 白色石膏板
22. 60毫米保温绝缘带
23. 预制混凝土板
24. 铝材与双层玻璃紧密贴合：层压安全玻璃+16毫米型腔 + 层压板安全玻璃
25. 2毫米钢包层漆框架
26. 钢结构
27. 3毫米铝板
28. 连续照明设备
29. 铝合金窗框双层玻璃：层压复古手绘安全玻璃 + 型腔+安全玻璃夹层
30. 由大理石砾岩瓷砖构成的浮动地板，尺寸600x600毫米，厚度30毫米
31. 有色铝管剖面，125x50毫米
32. 不锈钢护栏，排水沟
33. 不锈钢烤架
34. 外部混凝土路面
35. 混凝土门槛

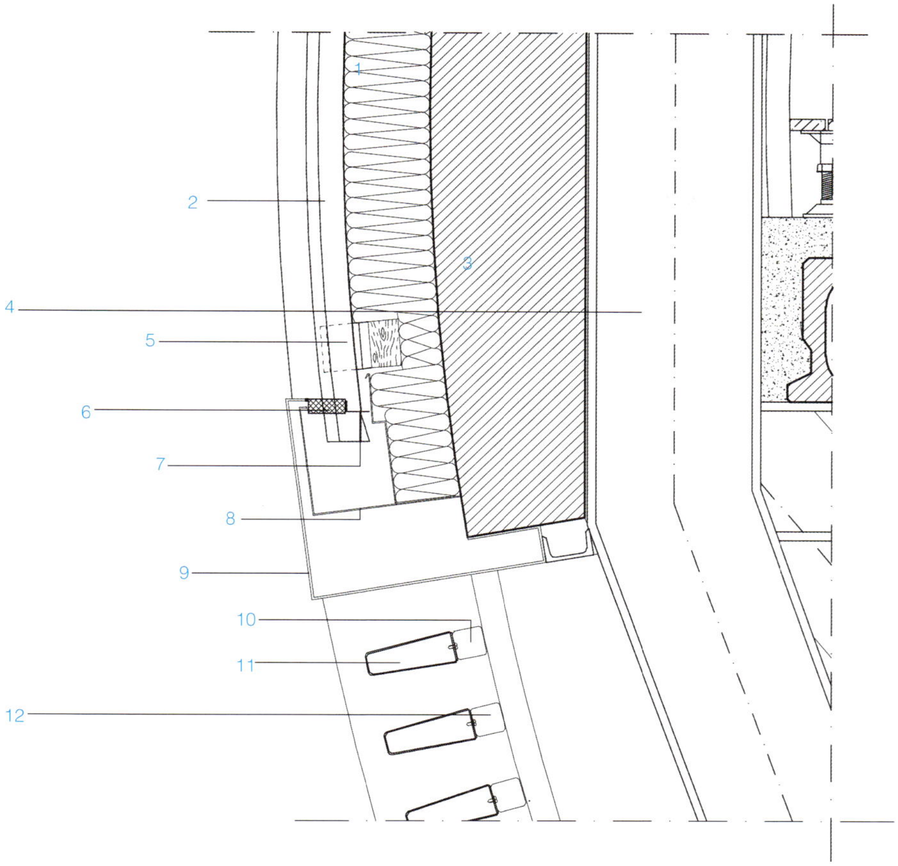

1. Mineral wool insulation thickness 140mm
2. Aluminium special cladding piece 65/400, 1mm
3. Pre-cast concrete element
4. IPE 80
5. Fixing clip for cladding
6. Polythene plug
7. Galvanized steel angle
8. Recessed steel drainage gutter
9. Aluminium cladding frame
10. Steel tubular section 50/50, 3mm
11. Aluminium section breaking-sun
12. UNP 80

1. 岩棉保温层厚度140毫米
2. 特殊的铝制包层片65/400，1毫米
3. 预制混凝土构件
4. 工字钢 IPE 80
5. 熔覆固定扣
6. 聚乙烯插头
7. 镀锌角钢
8. 嵌入式钢材排水沟
9. 铝包层框架
10. 50/50钢管截面，3毫米
11. 铝型材截面有效遮光
12. 液压油封规格 UNP 80

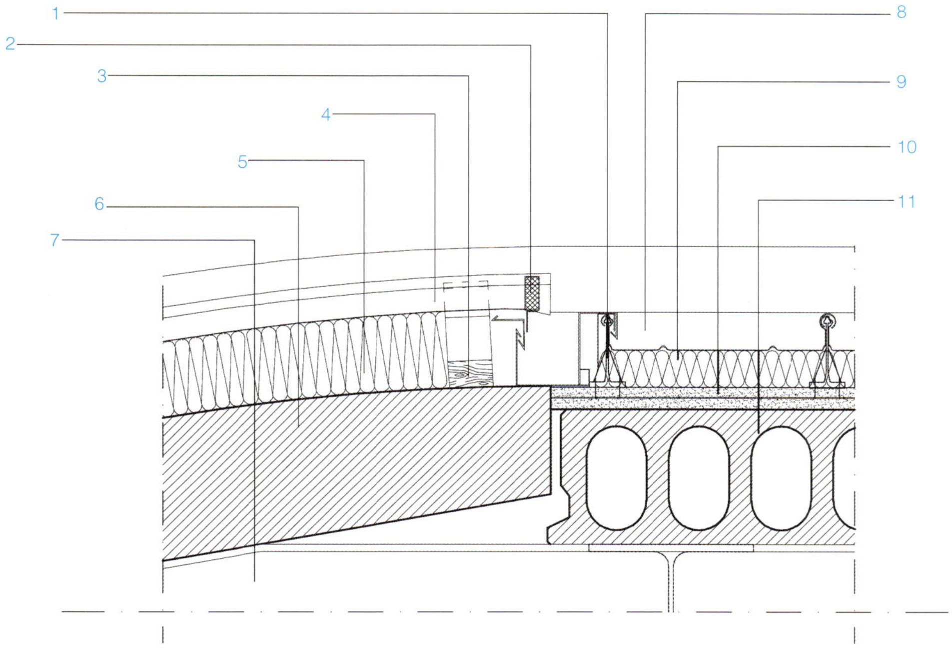

1. Aluminium fixing clip
2. Polythene plug
3. Pre-treated wood element
4. 65/400 aluminium cladding planks, thickness 1.0mm
5. Mineral wool insulation, thickness 140mm
6. Precast concrete element
7. Steel section IPE 80
8. 65/400 aluminium cladding planks, thickness 1.0mm
9. Mineral wool insulation, thickness 100 mm
10. Reinforced concrete layer pending 2.6%
11. 250 mm reinforced concrete slab

1. 铝材固定扣
2. 聚乙烯插头
3. 预先处理过的木材元素
4. 65/400铝包层木板，厚度1.0毫米
5. 矿棉绝缘板，厚度140毫米
6. 预制混凝土构件
7. IPE 80型钢
8. 65/400铝包层木板，厚度1.0毫米
9. 矿棉绝缘板，厚度100毫米
10. 钢筋混凝土悬层2.6%
11. 250毫米的钢筋混凝土板

Advice House

咨询大厦

Location: Vejle, Denmark
Designers: C. F. Møller Architects
Photographer: Julian Weyer
Completion date: 2009

项目地点：丹麦，瓦埃勒
设计师：C. F. Møller建筑师事务所
摄影师：朱利安・韦耶
竣工时间：2009年

Advice House is the first completed building in the Lysholt Park, a new business-park in the north of Vejle, and is with its proximity to the motorway designated to act as landmark and eye-catcher for the entire development. C. F. Møller Architects have developed two office-projects, Advice House and Lysholt Tower, for the client Lysholt Erhverv A/S. Both projects employ a simple, yet visually strong cladding with an unusual, colour-changing appearance.

The Advice House interior is 5,000 square metres of open and flexible office layout, where various tenants share the same large space, which offers dramatic perspectives and angles. The building is shaped around two angled office wings, separated by an equally angled atrium, resulting in a plan resembling a hexagon with one angle pushed inwards. The two wings are connected by walkways across the atrium, and the floors' continuous window-bands give a high degree of freedom in the space-planning. A large, north facing glazed gable gives passers-by a glimpse into the dynamic void, day or night, and the open and transparent interior is also naturally ventilated.

The building's unusual geometry makes for a dramatic and changing appearance when driving by on the motorway, and this mutability in form and shadows is further heightened by the colouring and texturing of the façades, designed to catch the light. The cladding-strips are composed of an "andom" sequence of a total of 13 differently proportioned cladding panels, some of which are folded diagonally to create a triangulated pattern. The panels are mounted horizontally at staggered intervals, creating a glittering array of colours, light and shadows.

The cladding panels are made from aluminium with a special colour pigmentation that offers changing colour effects with highlights and interesting colour gradients, depending on the viewing angle and the angle of the sun. Thus, the building never appears in quite the same way, and the effect is especially striking when passing by on the motorway.

咨询大厦是荷兰瓦埃勒市北部的莱索特商业园首个竣工建筑。因靠近公路，因此是整个规划项目中的地标性建筑，十分引人注目。C. F. Møller事务所设计了两座办公楼，一座是咨询楼，一座是莱索特大厦。两座建筑凭借简约的外观及独特、色彩变幻的表皮能够轻松捕捉人们的目光。

咨询大厦内部设有占地5,000平方米的开放而灵活的办公空间。不同的承租人分享同一个大型空间，形成了独特的视角。两座办公辅楼以角度相对构成了整座建筑，并以一个同角度的中庭为分界，形成了一个弯角向内的六边形。穿越中庭的通道连接了两座辅楼，连续的窗户为空间规划带来了高度自由性。大型北向的玻璃山形墙隐约地将活跃的室内气氛呈现给室外的行人，并有效地增强室内自然采光和通风效果。

建筑独一无二的几何形状打造了一个千变万化的立面，而独特的色彩和纹理更加强调了这种变幻效果。13种不同比例的外墙板"随意"构成了立面的条带，部分条带还对角折叠，形成三角形图案。外墙板垂直交叉设计，形成了光影色彩的闪亮阵列。

铝制外墙板，采用了特殊的色彩涂层，并根据观看和太阳照射角度的不同而打造出独特的色彩效果。这座美轮美奂、色彩斑斓的建筑在车流交织的公路旁尤为抢眼。

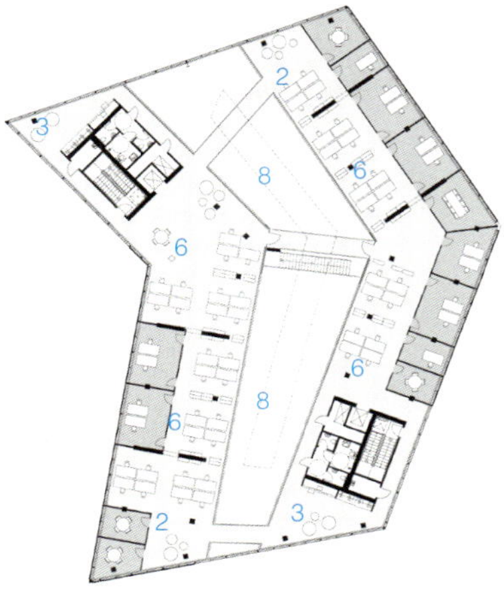

1. Entrance	1. 入口
2. Meeting	2. 会议室
3. Lounge	3. 休息室
4. Atrium	4. 中庭
5. Reception	5. 接待处
6. Office	6. 办公间
7. Canteen	7. 食堂
8. Atrium void	8. 中庭空间

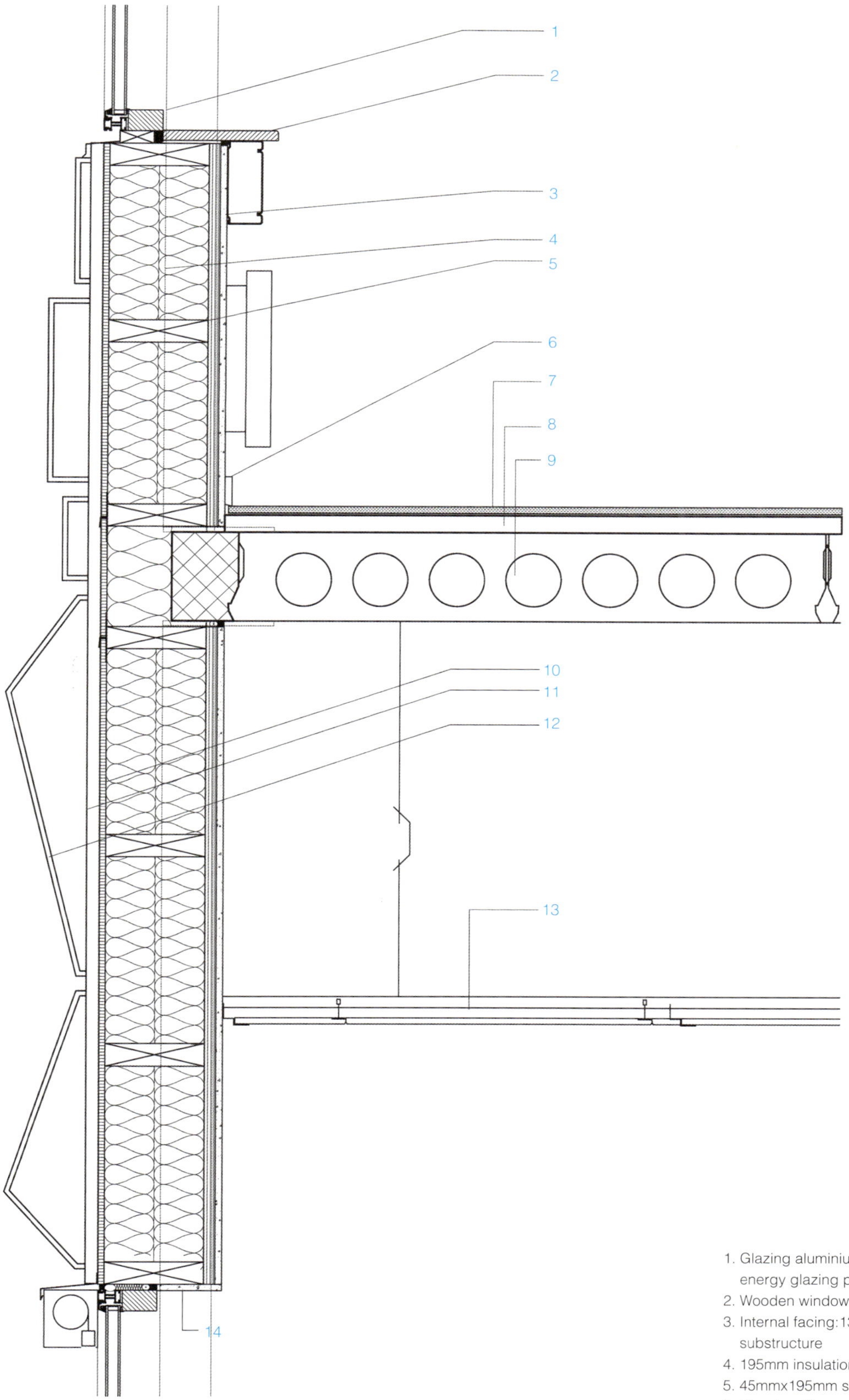

1. Glazing aluminium/wood frame system low-energy glazing panels
2. Wooden window sill
3. Internal facing:13mm plasterboard 22mm substructure
4. 195mm insulation
5. 45mmx195mm substructure
6. Wood profile
7. 14mm wood floor
8. 35mm anhydrite
9. Pre-fab concrete slab
10. 12mm fibre-cement board
11. Omega-section substructure
12. Façade: aluminium composite panels flat or folded spectra-colour coating
13. Suspended steel ceiling
14. 15mm plasterboard

1. 压光铝/木框架系统低耗能玻璃板
2. 木制窗台板
3. 内部饰面：13毫米厚石膏板，22毫米厚底座
4. 195毫米厚绝缘板
5. 45x195毫米厚基座
6. 木质剖面
7. 14毫米厚实木地板
8. 35毫米厚硬石膏
9. 预制纤维混凝土板
10. 12毫米厚纤维水泥板
11. 欧米加高硅型材底座
12. 外观：铝制复合平板或折叠光谱彩色涂层
13. 钢梁悬垂天花板
14. 15毫米厚石膏板

Seewurfel, Mixed-Use Regeneration Development

湖水立方体多功能再生工程

Location: Zurich, Switzerland
Designers: Camenzind Evolution
Completion date: 2005

项目地点：瑞士，苏黎世
设计师：Camenzind Evolution建筑师事务所
竣工时间：2005年

The eight new apartment and office buildings are situated close to the town centre of Zurich and offer stunning views of the lake and surrounding cityscape. The project regenerates a former industrial site into a new attractive centre for working and living and integrates itself harmoniously into the existing historic fabric of the area. The project was awarded the RIBA World-Wide Award 2005 by the Royal Institute of British Architects.

The project Seewurfel (meaning "Lake Cubes") is based on a concept of piazzas that were created by the careful positioning of the eight buildings. The exceptional quality of the project lies in the balance achieved between the unique modern architectural language of the development, the individual identity given to buildings by avoiding uniformity and the harmonious way in which the project integrates into the historic fabric of the neighbourhood.

This was achieved by creating buildings that were designed to be different in shape and size, while applying the same principles to details such as windows, cladding, entrances and staircases.

The concept of materials for the cladding develops further the principles of individuality and integration. The grey fibre-cement cladding panels form the unifying element that was used on all buildings. To give the buildings their own individuality and unique identity within the Seewurfel development, Camenzind Evolution specifically developed a new silicon-bonded timber-glass-panel cladding system. The 20 millimetres thick panels consist of two layers of 6 millimetres semi-toughened glass and an 8 millimetres veneered and UV-varnished MDF board. The three different types of wood-veneer used are from certified managed-forest sources and were selected for their individual colour and grain which give each building its distinctive character. The strong colour and texture of the timber-glass panels, together with the soft, moving reflections on the glass, create a human ambience of lightness and warmth in the piazzas between the buildings.

The Seewurfel development complies with the highest Swiss standards for energy efficiency (Minergie) and uses a geothermal heat pump system for environmentally-friendly heating and cooling.

湖水立方体多功能再生工程是集8个公寓和办公建筑于一体的小型社区，紧邻苏黎世市中心，其塑造的湖面景色和细腻、独特的设计为该区域增添了一道亮丽的城市景观。该工程将原有的工业厂区转变为一个富有吸引力的多功能区，同时和谐地融入到周边现存的历史街区和城市肌理之中，该项目也因此在2005年获得了由英国皇家建筑师协会颁发的世界建筑奖。

Seewurfel工程（寓意湖水立方体）的设计理念以八幢建筑的巧妙定位为基础，这八幢建筑的完美衔接又打造了一系列室外庭院，从而为空间的主人及其邻居提供风格各异的场景和不同的外部空间感受。该项目的独特之处在于运用独到的现代建筑设计语言，赋予建筑以自身的个性特征，并使二者之间的关系得以均衡，避免了与周边建筑在单纯意义上的重复，同时又以一种和谐的方式整合到现有的邻里环境和脉络当中。该设计理念的实现是通过遵循同一原则下的细部设计，例如开窗、覆层、入口、楼梯等，运用不同的形状和尺度来实现变化和统一。

建筑表皮材料的选用遵循个性化和一体化原则。灰色的纤维水泥板覆层形成统一的元素，应用到八个建筑之中。为体现每个建筑的特性，Camenzind Evolution建筑师事务所专门开发出新的硅－木材－玻璃板综合覆层系统。20毫米厚的板材中包括两层6毫米厚的半钢化玻璃、8毫米厚胶合板及紫外线涂层中密度纤维板厚板。不同类型的板材均系优良木材，个性化色彩和纹理赋予每个建筑以独特的魅力和特征。强烈的色彩和木料－玻璃板材纹理与玻璃上柔和的反射成像一起，在建筑之间的室外庭院中营造出温暖、清新的氛围。

整个项目符合瑞士最高的节能标准，并使用了环保的地热热泵系统实现空间的加热和制冷。

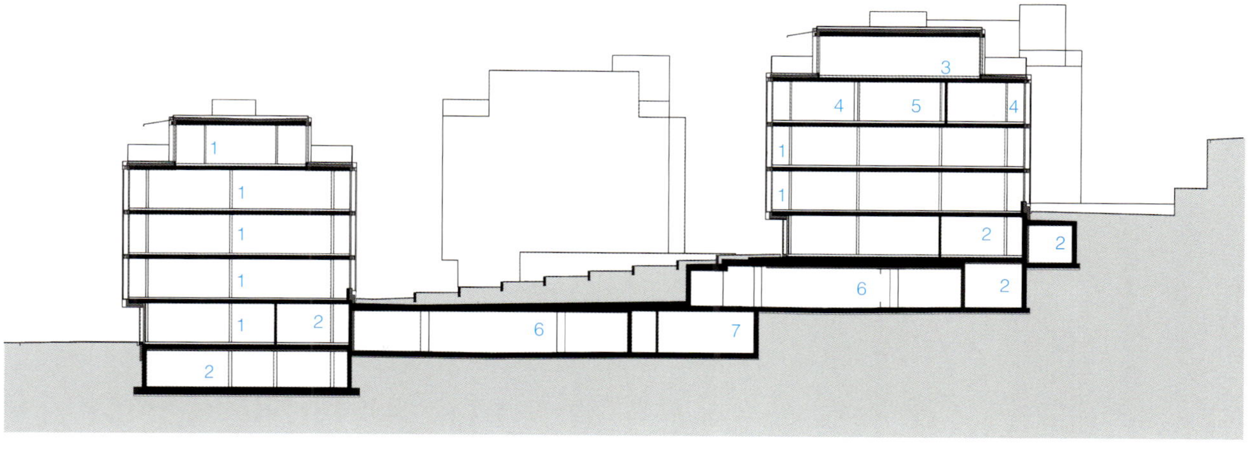

1. Office area
2. Lager storage
3. Living/dining
4. Zimmer room
5. Studio study
6. Underground parking
7. Plant room

1. 办公区
2. 啤酒储藏室
3. 客厅/饭厅
4. 房间
5. 研究工作室
6. 地下停车场
7. 设备间

1. Concrete/beton ceiling 250mm
2. Glass wool thermal insulation-dammung glaswolle 180mm
3. Ventilation gap
4. Aluminium silicon-bonded substructure-leminatedl aluminium-frame with studs on honrizontal supporting beam suspended
5. Timber-glass-panel/holzglaspaneel 20mm-semi-toughend glass/TVG 6mm-MDF-board veneered/MDF-platte mit furnier 8mm-veneer/furnier makore pommele, curubixa, bambus-semi-toughend glass/TVG 6mm
6. Lift and slide window
7. Aluminium external venetian blinds
8. Glass slider
9. Aluminium cladding casing
10. Guide wheels and guide rail substructure

1. 混凝土/ 天花板250毫米
2. 玻璃棉隔热，格拉斯沃莱玻璃纤维 180毫米
3. 通风间隙
4. 铝硅结合底部构造，粘合铝制框架以螺栓固定在平行支撑架上
5. 木材—玻璃板/木玻璃板20毫米—半透明玻璃/6毫米中密度纤维板/中密度纤 维板固定单板，8毫米胶合板/单板球纹猴子果木，竹子半透明玻璃/TVG板6毫米
6. 提升推拉窗
7. 外部铝制百叶帘
8. 滑动玻璃窗
9. 铝包层套管
10. 导轮与导轨底座

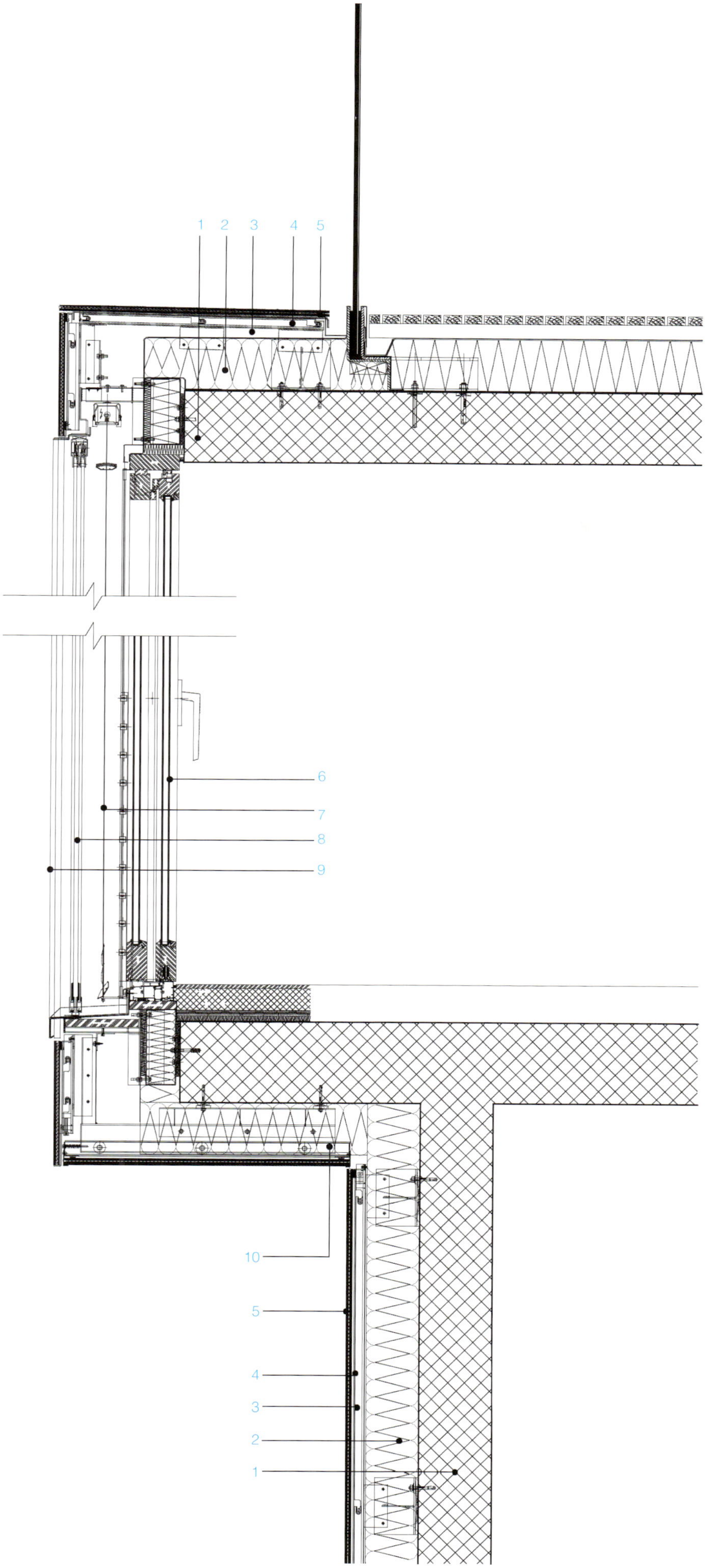

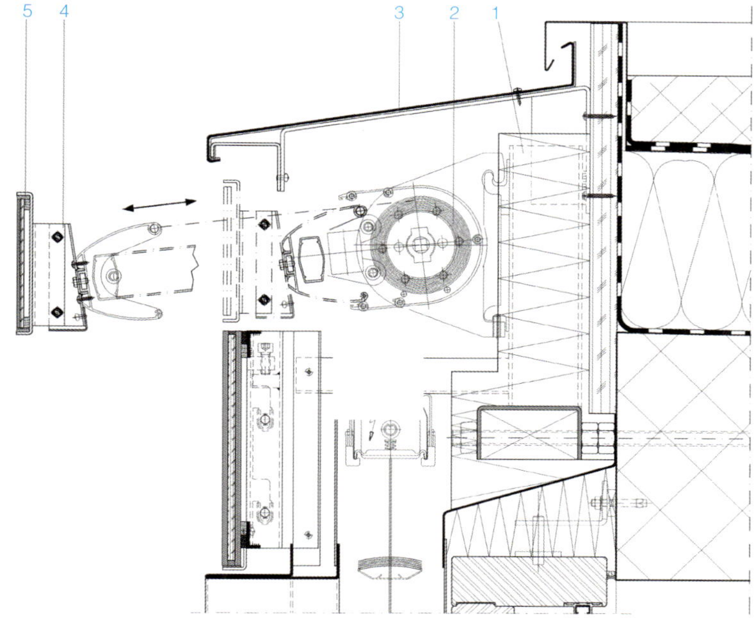

1. Substructure brackets
2. Folding arm awnings
3. Aluminium cladding
4. Substructure for the glass-timer-panel with adjustment features
5. Glass-timber-panel with rear aluminium sheet

1. 底层结构支架
2. 折臂遮阳篷
3. 铝板
4. 具有调整功能的玻璃板定时器下部结构
5. 背面设有铝板的玻璃木板

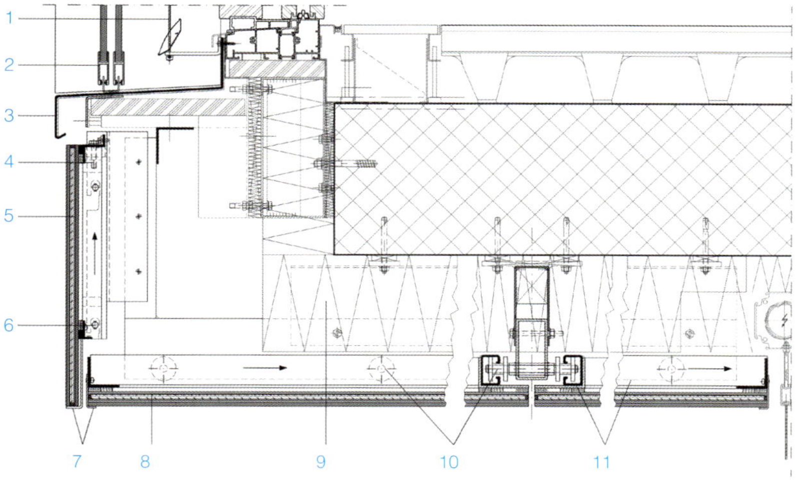

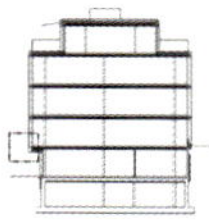

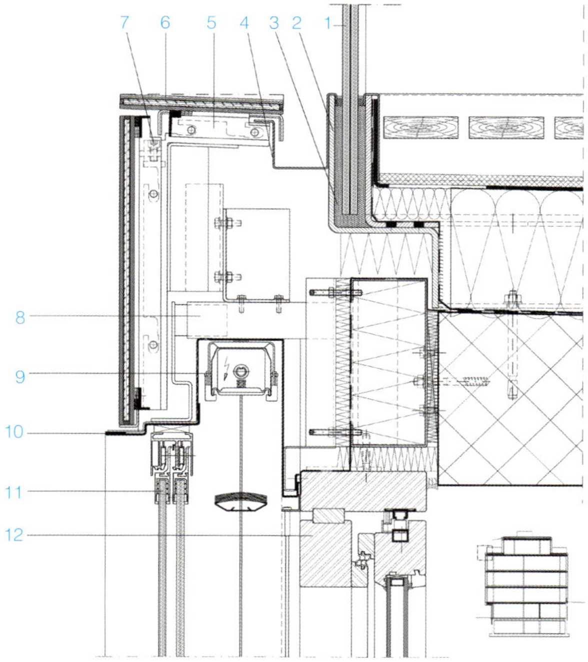

1. External venetian blinds
2. Glass slider
3. Aluminium window sill
4. Screw to adjust elevation and inclination
5. Glass-timber-panel 20mm; semi-toughend glass 6mm; MDF-board veneered 8mm; veneer-semi-toughend glass 6mm
6. Mounting pin
7. Stainless steel glass support bracket
8. Glass-timber-panel
9. Substructure brackets
10. Guide wheels
11. Guide rail

1. 外部活动百叶窗
2. 玻璃滑块
3. 铝合金窗台板
4. 螺丝调整高度和倾角
5. 玻璃木板20毫米；半钢化玻璃6毫米；8毫米胶合板中密度纤维板；胶合板－半钢化玻璃6毫米
6. 安置销
7. 不锈钢玻璃支架
8. 玻璃木材板
9. 底层结构支架
10. 导向轮
11. 导轨

1. Structural glass balustrade
2. Steel substructure
3. Structural resin filling
4. Stainless steel gutter
5. Aluminium substructure
6. Glass-timber-panel
7. Screw to adjust elevation and inclination
8. Window and venetian blinds substructure
9. External venetian blinds
10. Aluminium cladding frame
11. Glass slider
12. Timber sliding windows

1. 结构玻璃栏杆
2. 钢架结构
3. 树脂充填结构
4. 不锈钢水槽
5. 铝底座
6. 玻璃木材板
7. 螺丝调整高度和倾角
8. 窗口与活动百叶窗
9. 外部百叶窗
10. 铝包层框架
11. 玻璃滑块
12. 木质滑动窗口

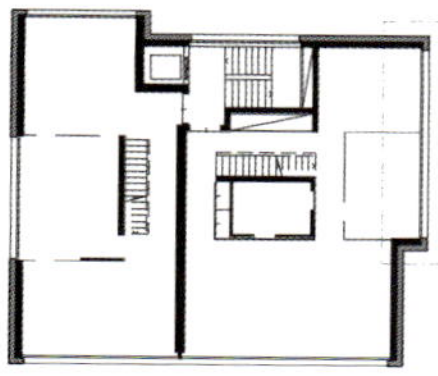

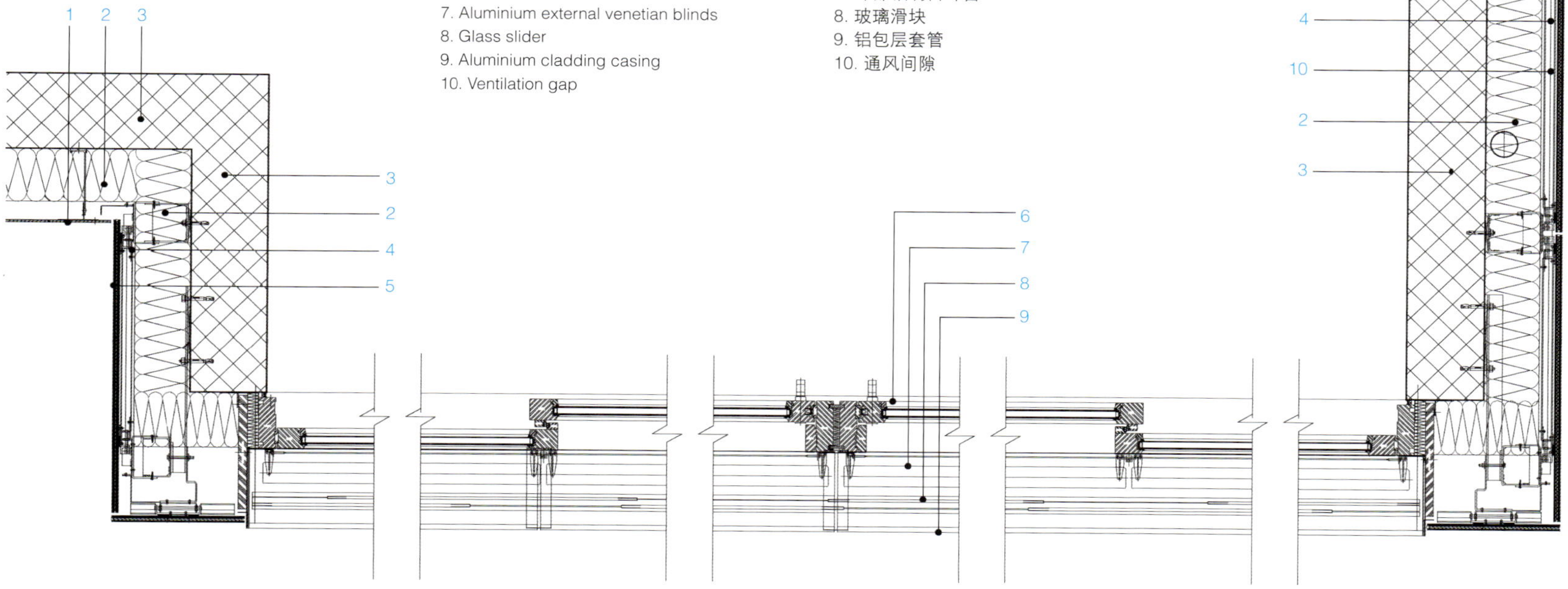

1. Fibre-cement panel 8mm
2. Glass wool thermal insulantion 180mm
3. Concrete 250mm
4. Aluminium silicon-bonded substructure
5. Glass-timber-panel 20mm; semi-toughend glass 6mm; MDF-board veneered 8mm; veneer-semi-toughend glass 6mm
6. Timber-sliding windows
7. Aluminium external venetian blinds
8. Glass slider
9. Aluminium cladding casing
10. Ventilation gap

1. 8毫米纤维水泥板
2. 180毫米隔热玻璃丝
3. 250毫米厚混凝土
4. 铝硅结合底层结构
5. 玻璃木板20毫米；半钢化玻璃6毫米；8毫米胶合板中密度纤维板；胶合板－半钢化玻璃6毫米
6. 木质滑动窗口
7. 外部铝制百叶窗
8. 玻璃滑块
9. 铝包层套管
10. 通风间隙

Index 索引

1. Dissing+Weitling Architecture
Contact: SMI@dw.dk

2. Fentress Architects
Contact: Popowski@fentressarchitects.com

3. Cepezed Architects
Contact: info@cepezed.nl

4. GDP Architects Sdn Bhd
Contact: tuti@gdparchitects.com

5. Alsop
Contact: cskillings@strattonandreekie.com

6. patrick genard & asociados
Contact: dariela@patrickgenard.com

7. Ferdinand and Ferdinand Architects
Contact: ferdinand@nextra.hu

8. MONK & Paul Dam
Contact: info@monkpaul.com

9. Coll-Barreu Arquitectos / Juan Coll-Barreu / Daniel Gutiérrez Zarza
Contact: madrid@coll-barreu-arquitectos.com

10. Tony Owen Partners
Contact: info@ton.com

11. Archimedialab
Contact: bl@archimedialab.com

12. Rojkind Arquitectos
Contact: jakmak@club-internet.fr

13/32. Camenzind Evolution
Contact: info@camenzindevolution.com

14. Saucier+Perrotte Architects
Contact: design@sadarvuga.com

15. HG Arhitektuur
Contact: tomo@hga.ee

16. J. MAYER H. Architects
Contact: contact@jmayerh.de

17. Dietmar Feichtinger Architectes
Contact: contact.paris@feichtingerarchitectes.com

18. VAÍLLO & IRIGARAY + EGUINOA
Contact: estudio@vailloirigaray.com

19. Mossessian & Partners
Contact: theresa@theresasimon.com

20. gmp-von Gerkan, Marg and Partners Architects
Contact: bahrens@gmp-architekten.de

21. Dimitris Papaioannou + Associates Architects, Planners, Engineers
Contact: gdpap@tee.gr

22. Robert McBride, Debbie-Lyn Ryan
Contact: info@McClarles.com

23. barbosa & guimarães
Contact: info@boborsa.com

24. UArchitects (Eindhoven, The Netherlands)
Contact: emilevanvugt@uarchitects.com

25. MIII Architects
Contact: info@m3architecten.com

26. BOB361 Architecten
Contact: bxl@bob361.com

27. Steven Holl Architects
Contact: julia@stevenholl.com

28. Audrius Ambrasas Architects
Contact: Info <info@ambrasas.lt>

29. BDP
Contact: martin.davies@bdp.com

30. Massimo Mariani
Contact: info@massimo.com

31. C. F. Møller Architects
Contact: JW@cfmoller.com

图书在版编目（CIP）数据

建筑表皮细部结构 / 高巍编 ; 贺丽译. -- 沈阳：辽宁科学技术出版社，2011.10
ISBN 978-7-5381-6939-3

I．①建… II. ①高… ②贺… III. ①建筑物－外墙－细部设计 IV. ①TU227

中国版本图书馆CIP数据核字（2011）第076250号

出版发行：辽宁科学技术出版社
（地址：沈阳市和平区十一纬路29号 邮编：110003）
印 刷 者：利丰雅高印刷（深圳）有限公司
经 销 者：各地新华书店
幅面尺寸：245mm × 290mm
印 张：36
插 页：4
字 数：50千字
印 数：1~2000
出版时间：2011年 10月第 1 版
印刷时间：2011年 10月第 1 次印刷
责任编辑：陈慈良
封面设计：赵 聪
版式设计：赵 聪
责任校对：周 文

书 号：ISBN 978-7-5381-6939-3
定 价：268.00元

联系电话：024-23284360
邮购热线：024-23284502
E-mail: lnkjc@126.com
http://www.lnkj.com.cn
本书网址：www.lnkj.cn/uri.sh/6939